General Engineering Knowledge

General Engineering Knowledge

THIRD EDITION

H. D. McGeorge, CEng, FIMarE, MRINA

LONDON AND NEW YORK

First published by Stanford Maritime Ltd 1978
Subsequently published by Butterworth-Heinemann

This edition published 2011 by Routledge
2 Park Square, Milton Park, Abingdon, Oxon OX14 4RN
711 Third Avenue, New York, NY 10017, USA

Routledge is an imprint of the Taylor & Francis Group, an informa business

Notice
No responsibility is assumed by the publisher for any injury and/or damage to persons
or property as a matter of products liability, negligence or otherwise, or from any use
or operation of any methods, products, instructions or ideas contained in the material
herein. Because of rapid advances in the medical sciences, in particular, independent
verification of diagnoses and drug dosages should be made

British Library Cataloguing in Publication Data
McGeorge H. D. (H. David)
  General engineering knowledge – 3rd ed.
  1. Marine engineering
  I. Title
  623.87

Library of Congress Cataloging-in-Publication Data
A catalog record for this book is available from the Library of Congress
ISBN: 978-0-7506-0006-4

Transferred to Digital Printing in 2013

Contents

Preface

The written examinations in engineering knowledge for Merchant Navy Certificates of Competency Classes 1, 2 and 3 all have the same general content. There are, of course, differences of emphasis and in the way that the questions are asked. Information in this book is intended to be of assistance to candidates for all of the papers.

This third edition of **General Engineering Knowledge** has been expanded and updated to cover changes in the examination questions and legislation introduced since the previous edition. The chapter on pollution prevention now includes sections on disposal of chemicals and garbage, in addition to notes on prevention of pollution by oil, the Clean Air Act and disposal of sewage. A new chapter on production of water by low-pressure evaporators and reverse osmosis contains notes on treatment to make the water potable and on problems with bacteria.

Noise, another form of pollution, is also associated with vibration and there is now a chapter dealing with both topics. The section on vibration covers its use as a means of monitoring the condition of machinery.

Additions have been made to various chapters and references where appropriate for further reading.

'Knowledge is of two kinds — we know a subject ourselves, or we know where we can find information upon it.' (Sam Johnson)

H. D. McG.

CHAPTER 1

Centrifugal Pumps and Priming— Coolers and Cooling Systems— Pipelines and Corrosion

CENTRIFUGAL PUMPS

The simple centrifugal pump is used for sea water circulation and other duties where self priming is not a requirement. When installed for bilge pumping or ballast duty, these pumps require a primer i.e. some means of removing air from the suction pipe so that the liquid to be pumped is caused to flow into the pipe and so to the eye of the impeller.

SINGLE STAGE CENTRIFUGAL PUMP

For general duties the impeller is of aluminium bronze keyed and secured to a stainless steel shaft. The impeller shown (Fig. 1) is fully shrouded and of the

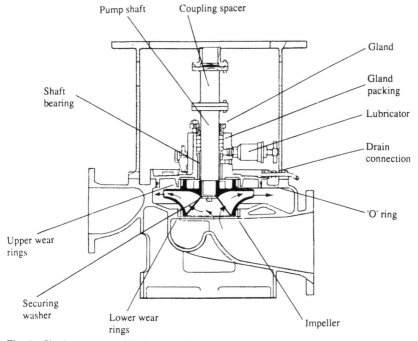

Fig. 1   Single stage centrifugal pump *(Hamworthy Engineering Ltd)*

single entry type. The renewable wear rings are of aluminium bronze and the casing is normally of bronze or cast iron. The cover has a hub containing the shaft bearing at the bottom and, above, either a packed gland or a mechanical seal. The shaft bearing is of phenolic resin asbestos, lubricated by the liquid being pumped except for pumps operating on high static lift. These have grease lubricated bronze bearings to ensure adequate lubrication during the priming period.

A spigotted coupling spacer connects the motor half coupling to the pump shaft. When this is removed, the pump cover, together with the impeller and shaft assembly can be lifted out of the pump casing for inspection or maintenance.

IMPELLERS

The fully shrouded, single entry impeller in the pump shown (Fig. 1) is the type most widely used. It consists of a number of vanes curving backwards from the direction of rotation. The vanes are supported on one side by shrouding connected to the hub. The shrouding supporting the vanes on the other side, has an entry at the centre. When the pump is operating, liquid in the casing is swirled by the rotating impeller. The swirling action causes the liquid to move towards the outside and away from the centre (in the same way that stirred coffee moves to the side of the cup, tending to spill over the rim and leaves a dip at the centre). The backward curving vanes and the rotation give the liquid a combined radial and circular motion.

CASING

The section of the volute casing shown in the sketch (Fig. 2) increases, thus allowing unrestricted flow from the impeller. The volute also acts as a diffuser,

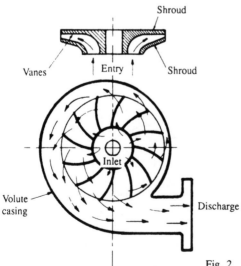

Fig. 2   Section through impeller and casing

converting kinetic head into pressure head. Some pumps have a double volute casing which gives radial balance and reduced wear on the bearings.

Pumps designed to produce high pressure, have a diffuser ring so that a greater quantity of kinetic energy in the liquid can be converted to pressure.

SUCTION

When a centrifugal pump is operating, the liquid leaving the impeller produces a drop in pressure at the entry or eye of the impeller. This causes liquid from the suction pipe to flow into the pump. In turn, there is a movement of the liquid to be pumped. The latter is normally subject to atmospheric pressure. A centrifugal pump will maintain a suction lift of four metres or more once it has been primed, because of the water passing through.

The water in a pump acts like a piston for water in the suction pipe and an empty pump will not operate.

A pump which is required to initiate suction from a liquid level below itself, must be fitted with an air pump.

AIR PUMP ARRANGEMENT

The diagram (Fig. 3) shows a primer coupled to the top of an electric motor and centrifugal pump set. A pipe from the pump outlet, provides cooling water for the primer. This returns through another pipe to the pump suction.

The main pump suction pipe has a float chamber fitted. The float operates a valve on the pipe leading from the float chamber to the air pump suction. With no liquid in the suction, the float drops, opening the valve and allowing the air pump to evacuate the air from the suction pipe. This partial vacuum causes the atmospheric pressure to force liquid into the suction pipe. The rising liquid will lift the float and close the valve on the air pump suction. Air pumped out, passes to atmosphere.

AIR PUMP PRINCIPLE

The air pump or water ring primer, as the simple plan view shows (Fig. 4) consists of an elliptical casing which contains a vaned rotor and has a covering plate with ports cut in it. The casing is partly filled with water. The rotor is coupled to the electric motor so that when the pump is running the water spins with the rotor and being thrown outwards, takes up an elliptical shape. The tips of the vanes are sealed by the water and the volume between them varies during the rotation. Beneath the suction ports, the volume increases so that air is drawn from the float chamber. Under the discharge ports, the volume decreases, forcing air out.

Cooling water is necessary to prevent overheating of the sealing water from the action of the vanes in the liquid. Interruption of the coolant supply results in vapour from the sealing water destroying the vacuum effect, so that air is no longer pumped.

The internal passages of a typical air pump are shown in the sectional sketch (Fig. 5). The right side shows the operating passages and the path of the air being pumped. It is drawn from the suction float chamber of the main pump and through the pipe and passages to the suction ports of the primer. The discharge

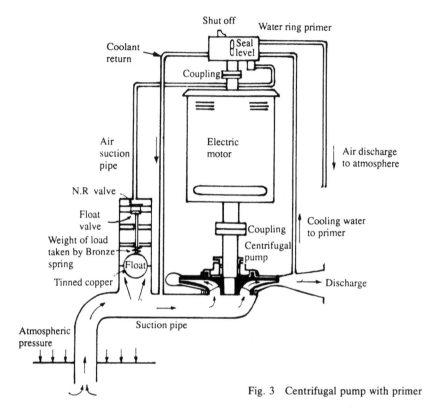

Fig. 3   Centrifugal pump with primer

ports are not shown but the air from them is discharged into the top of the outer casing. From the outer casing the air is discharged to atmosphere.

The primer runs continuously but can be unloaded, when the pump has been primed, by the arrangement on the left. The shut-off handle rotates a ported

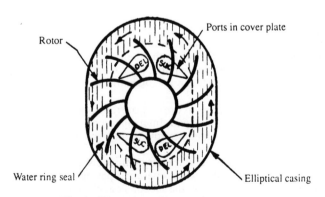

Fig. 4   Water ring primer or air pump

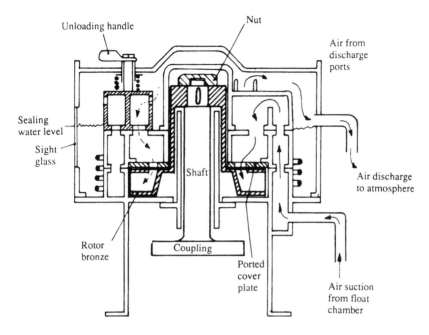

Fig. 5   Air pump showing operating passages

chamber to the position shown, so that the partial vacuum is broken and the water is free to circulate.

OPERATION

Centrifugal pumps for bilge, ballast and general service are usually fitted with primers. Before starting such pumps, the primers must be checked, to ensure that the sealing water is at the correct level. Fresh water is used for topping up. Suction valves between the liquid and the pump are opened and a check is made that other valves on the suction side of the system are closed. The delivery valve is kept shut and the pump is started. The centrifugal pump can be started with the discharge valve closed, but it is an exception. (A propeller pump should not be started with the discharge shut or overload results. This is shown by comparison of characteristic curves. Obviously positive displacement pumps would not be started with closed discharges.) If priming takes a long time, the primer will become hot unless cooling water is passed through it. The sea water suction can be opened to allow cooling.

EXHAUSTER FOR CENTRAL PRIMING SYSTEM

Several centrifugal pumps can be primed from a central vacuum tank as an alternative to being fitted with individual water ring primers. The pumps are connected to the vacuum tank through the same sort of float chamber arrange-

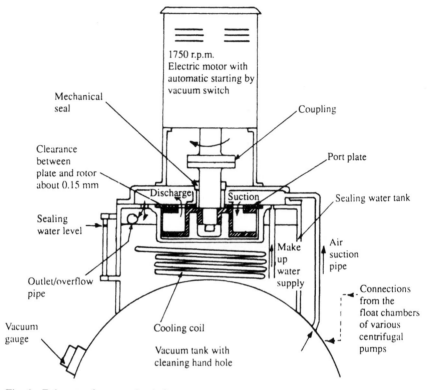

Fig. 6   Exhauster for central priming system

ment as is used with the individual water ring primer. There is also a shut off cock for isolating the pump and a non-return valve.

The primer has two electrically driven air pumps (Fig. 6) which evacuate the tank. Starting is by means of pressure switches through suitable starters, and air pumps are automatically stopped by the switches when the required vacuum is reached. The pumps run intermittently as demand makes necessary and not continuously. Thus, on a vessel where priming is needed for a number of pumps, the use of a central primer would reduce the number of air pumps and the running time for them.

The primer shown, consists of a gunmetal casing of oval shape with a rotor and ported plate, as described for the individual water ring primer. Water in the casing forms a seal and, because it takes up the elliptical shape of the casing when the rotor turns, produces a pumping action. The two suction ports are connected by passages in the cover to the suction pipe from the vacuum tank. The two discharge ports are connected via an aperture to the sealing water tank. Air from the vacuum tank, together with make-up water, is drawn into the suction and discharged to the sealing water tank. The water remains in the tank, while the air passes to atmosphere through the outlet/overflow pipe.

The sealing water reservoir also keeps the air pump cool and is cooled in turn, by a sea water flow through the cooling coils shown. If the water ring temperature rises, then the function of the primer will be destroyed by the presence of water vapour.

TUBE COOLERS

Tube coolers for engine jacket water and lubricating oil cooling are normally circulated with sea water. The sea water is in contact with the inside of the tubes and the water boxes at the cooler ends. Two pass flow is shown in the diagram (Fig. 7) but straight flow is common in small coolers.

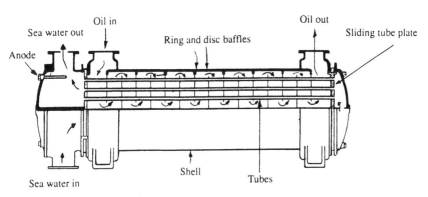

Fig. 7   Tube cooler with two pass flow (Serck)

The oil or water being cooled is in contact with the outside of the tubes and the shell of the cooler. Baffles direct the liquid across the tubes as it flows through the cooler. The baffles also support the tubes.

Tubes of aluminium brass (76 per cent copper; 22 per cent zinc; 2 per cent aluminium) are commonly used. Ordinary brasses and other cheap materials have been used with unsatisfactory results. The successful use of aluminium brass has apparently depended on the presence of a protective film formed along the tube length by corrosion of iron in the system. Thus unprotected iron in water boxes and other parts, while itself corroding, has prolonged tube life. This was made apparent when steel was replaced by other corrosion resistant materials or protected more completely. The remedy in these systems has been to fit sacrificial soft iron or mild steel anodes in water boxes or to introduce iron in the form of ferrous sulphate fed into the sea water. The latter treatment consists of dosing the sea water to a strength of 1 ppm for an hour per day over a few weeks and subsequently to dose before entering and after leaving port for a short period.

Early tube failures may be due to pollution in coastal waters or to turbulence in some cases.

Many coolers are fitted with tubes of 70/30 cupro-nickel. More expensive materials are available. Tubes are expanded into tube plates and may be further bonded by soldering.

7

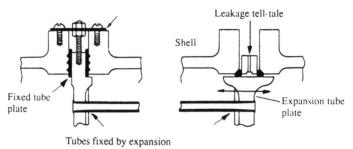

Fig. 8   Detail of cooler expansion arrangement

TUBE PLATES

Naval brass tubeplates are used with aluminium brass tubes. Tube stacks are made up to have a fixed tube plate at one end and a tube plate at the other end which is free to move with the expansion of the tubes (Fig. 8). Other materials found in service are gunmetal, aluminium bronze and sometimes special alloys.

TUBE STACK

The tube stack shown is fitted with alternate disc and ring baffles. The fixed end tube plate is sandwiched between the casing and the water box. If the joints leak at the other end the special 'tell tale' ring will allow the liquids to escape without mixing. The joint rings are of synthetic rubber.

WATER BOXES AND COVERS

Easily removable covers on water boxes permit repairs and cleaning to be carried out. The covers and water boxes are commonly of cast iron or fabricated from mild steel. Where they have been coated with rubber or a bitumastic type coating, the iron or steel has been protected but has provided no protection for the tubes and tubeplate. Uncoated ferrous (iron) materials in water boxes provide a protective film on the tubes as the unprotected iron itself corrodes, the products of corrosion coating the tubes. The iron also gives some measure of cathodic protection.

Water boxes of gunmetal and other materials are used but these, like the coated ferrous metals, give no protection. Soft iron or mild steel anodes can be fitted in the water boxes and provided they cause no turbulence, will help to give cathodic protection and a protective film.

SHELL

The shell or cylinder is fabricated or cast. It is in contact with the liquid being cooled. This may be oil, with which there is no corrosion problem, or water, which is normally inhibited against corrosion. The material is not critical (provided it is not reactive with any inhibiting chemicals) because it is not in contact with sea water.

INSTALLATION

Manufacturers recommend that coolers are arranged vertically. If horizontal installation is necessary the sea water should enter at the bottom and leave at the top. Air in the system will encourage corrosion and air locks will reduce the cooling area and cause overheating. Thus vent cocks should be fitted for purging air. Clearance is required at the cooler fixed end for removal of the tube nest.

SEA WATER SIDE

Only the minimum of salt water should be circulated in coolers. Thus it is best to regulate temperature by means of the salt water outlet valve, the inlet being left full open. If temperature is maintained by adjustment of the oil or jacket water flow, with full flow on the sea water side, there is a greater corrosion risk.

Strainers on the sea water pump suctions should be cleaned and checked regularly, as blockage will starve the system of water. Damage to the strainer plate will allow solids through which will block the end of the cooler. The cooler will become ineffective in either case and partial blocking of the cooler tends to lead to erosion damage.

The sea water side should be disturbed only when necessary to avoid damage to the protective film on the inside of the tubes. If cleaning is needed to remove deposits, use should be made of the special soft brushes. Chemical cleaning may be recommended particularly where hard deposits have accumulated. The manufacturers handbook will list acceptable cleaning chemicals. For the sea water side of coolers, an acid such as hydrochloric acid may be the agent.

Precautions are essential when dealing with corrosive chemicals used for cleaning. Contact is avoided by wearing gloves and protective goggles or a face shield. Should the chemical come into contact with the skin or the eye, the best first aid is usually to wash the affected parts immediately with water. If other treatment is necessary this can be found from the medical book. Before handling any chemical the instructions should be read and the type of first aid that might be necessary ascertained. There are now such a variety of chemicals in use that reference books are needed. Mixing instructions must be followed.

Before cleaning, coolers are isolated from the system by valves and blanks or by removing pipes and blanking the cooler flanges. Flushing is necessary after the cleaning agent has been drained from the cooler.

PLATE TYPE HEAT EXCHANGER

Plate type heat exchangers were originally developed for the milk industry where daily cleaning is necessary. They were first used at sea, as coolers, in the nineteen-fifties.

The plates are metal pressings (Fig. 9), corrugated with horizontal or chevron pattern corrugations. These make the plates stiffer and therefore permit the use of thinner material. They also increase the heat exchange area and produce a turbulent flow. All these factors contribute to the efficiency of heat transfer. Turbulence, as opposed to smooth flow, causes more of the liquid passing between the plates to come into contact with them. It also breaks up the boundary layer of liquid which adheres to the metal and acts as a heat barrier in smooth flow. However, the turbulence can cause plate damage due to erosion

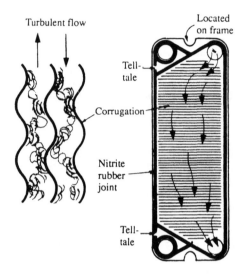

Turbulent flow

Located on frame

Tell-tale

Corrugation

Nitrite rubber joint

Tell-tale

Fig. 9   Cooler plate

and materials normally used in tube coolers for sea water contact, may not be suitable in plate coolers.

Plate material for sea water contact is titanium. This is an expensive metal but apparently able to withstand the conditions of service. Aluminium-brass has been used with poor results. Possibly failure of aluminium-brass has been due to the presence of organic sulphides and other chemicals in coastal and inland waters. (Titanium is immune from this type of attack.) However, other factors such as the turbulence in plate coolers or changes in the materials of sea water

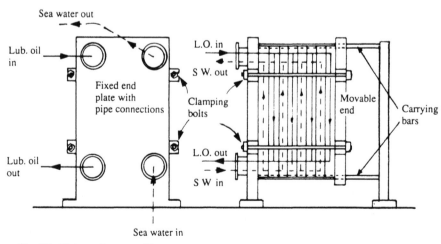

Sea water out

Lub. oil in

L.O. in

S W. out

Fixed end plate with pipe connections

Clamping bolts

Movable end

Carrying bars

Lub. oil out

L.O. out

S W in

Sea water in

Fig. 10   Plate cooler assembly

systems may be responsible for early failures. Stainless steel has been used in plate coolers for duty with sea water, but proved unsuccessful although it is suitable for other applications.

The joint material is normally nitrile rubber which is bonded to the plate with suitable adhesive such as Plibond. Other joint materials for higher temperatures are available, such as compressed asbestos fibre. The nitrile rubber is suitable for temperatures up to about 100°C (230°F). At high temperatures the rubber hardens and loses its elasticity. The rubber joints are compressed when the cooler is assembled and the clamping bolts tightened. Overtightening can cause damage to the chevron corrugated plates so the cooler stack must be tightened, and dimensions checked, during the process. Joints must be adequately clamped to prevent leakage.

All liquid inlets and outlets are at the fixed end plate. The movable end sits in the horizontal carrying bars and the plates are also located and supported by these. The flow ports at the corners of the plates are arranged so that the cooling liquid and the liquid being cooled pass between alternate pairs of plates. The sketch (Fig. 10) illustrates the way in which the liquids flow. Best efficiency is obtained by liquids moving in opposite directions i.e. contra-flow. Joint leakage is visible externally except for the double joint at the ports on one side of the plate. A drain hole acts as a tell-tale for this section (see Fig. 9).

ADVANTAGES AND DISADVANTAGES

Plate coolers are smaller and lighter than a tube cooler giving the same performance. No extra space is needed for dismantling (a tube cooler requires enough clearance at one end to remove the tube nest). Their higher efficiency is shown by the smaller size. Plates can be added, in pairs, to increase capacity and similarly damaged plates are easily removed, if necessary without replacement. Cleaning is simple as is maintenance. Turbulent flow helps to reduce deposits which would interfere with heat flow.

In comparison with tube coolers, in which tube leaks are easily located and plugged, leaks in plates are sometimes difficult to find because the plates cannot be pressurized and inspected with the same ease as tube coolers. Deteriorating joints are also a problem; they may be difficult to remove and there are sometimes problems with bonding new joints. Tube coolers may be preferred for lubricating oil cooling because of the pressure differential. Cost is another drawback; there are a large number of expensive joints on plate coolers and the plates are expensive.

METHODS OF SERVICING

The difficulty of removing old gaskets is overcome in factory servicing by the use of a liquid nitrogen spray. The plates are passed through a chamber containing the spray, on a conveyor belt. The intense cold makes the gasket brittle and as the metal of the plate contracts the resultant stresses set up between gasket and plate cause the glued joint to fail. It is sufficient to bang the plates once or twice after cooling to remove the joint debris.

Plates are cleaned before joint removal so that they are ready for crack detection afterwards. For this, they are sprayed with dye penetrant and viewed under an ultra-violet light to show up any defects. New joints are fitted using a thermo setting adhesive which is cured in an oven.

TITANIUM

Over the last thirty-five years, the corrosion resistance of titanium and its alloys has led to its development for use in sea water systems. Its light weight (density 4.5 kg/m^3) and good strength make it a useful material, but the main benefit is its corrosion resistance in static or fast flow conditions. It has a tolerance to high flow velocity which is better than that of cupro-nickel; it is also resistant to sulphide pollution in sea water. The modulus of elasticity of titanium is about half that of steel.

While titanium has great corrosion resistance itself because it is more noble than most materials used in marine systems, it also tends to set up galvanic cells with other metals. The cathodic titanium makes the other materials anodic and likely to suffer wastage. The possibility of corrosion is reduced by careful choice of compatible metals, coating of the titanium, insulation or use of cathodic protection.

PIPELINES AND CORROSION

Sea water pipes for circulation of cooling water, together with those for bilge and ballast systems, are prone to internal wastage from corrosion and erosion. External corrosion of steel pipes is also a problem in the tank top area.

SEA WATER PIPELINES

Ship side valves for sea water inlet must be of steel or other ductile material. The alternative materials are bronze, spheroidal graphite cast iron, meehanite or another high-quality cast iron. Ordinary grey cast iron has proved to be unreliable and likely to fail should there be shock from impact or other cause. Permissible cast irons must be to specification and obtained from an approved manufacturer.

Bronze has good resistance to corrosion but is expensive and therefore tends to be used for smaller ship side valves. Steel is cheaper, but prone to corrosion. It may be cast or fabricated. Unprotected steel valve casings and pipes will in the presence of sea water and bronze seats, valve lids and spindles waste due to galvanic corrosion. The presence of corroding iron or steel confers benefits on sea water systems. The metal acts as a sacrificial anode and additionally delivers iron ions which are carried through and give protection to other parts of the system where they deposit.

Wastage of steel will result where there is weld spatter and where there are differences between weld material and the steel of the pipe.

Pipes are fabricated and bent to shape before removal of weld spatter and scouring as a prelude to galvanizing both internally and on the outside (weld spatter which projects through the zinc of the galvanized protective layer, will cause concentrated corrosion).

Water speeds for galvanized steel pipes should be limited to 3 metres per second to avoid erosion. Thus, pipes should be of generous size, with bends shallow rather than sharp, no changes of section and no inward projection of joints.

Galvanic corrosion occurs in the presence of sea water, where there are different metals in the make up of a pipe system or differences in one material. In galvanic cells, the least noble metal wastes leaving the more noble intact. This gives the impression that more noble materials such as the copper and copper alloys are resistant to corrosion in sea water. Particular types of corrosion such as de-zincification do in fact cause problems in pipework and coolers. Concentrated galvanic corrosion occurs where there is a mix of non-ferrous materials. De-zincification is the loss of zinc from a brass with the result that a soft spongy copper is left. The fault leads to leakage around tubes in tube plates so affected and other problems. Areas which have suffered from this particular form of corrosion are soft and of a red copper colour. De-zincification is inhibited in brasses for marine use, by additions of minute quantities of elements such as arsenic (0.04%) or other alloy materials.

Non-ferrous pipe and cooler systems have suffered from unusual problems as the result of the mix of different copper alloys, the lack of iron ions from corroding iron or steel and possibly from high water speeds.

Recommendations have been made for the installation of steel sacrificial anodes in cooler water boxes, use of thick unprotected steel cooler covers, fitting of steel pipe sections or recourse to driven steel anodes to provide protective iron ions and galvanic protection. Injections of ferrous sulphate are also resorted to as a means of supplying beneficial iron ions for non-ferrous systems. Protection is provided by impressed current through platinized titanium anodes in some cooling systems. Plastic inserts are inserted at the inlet ends of some cooler tubes to reduce otherwise rapid erosion (and corrosion) in a vulnerable zone.

Water speeds in non-ferrous as in galvanized steel pipes, must be limited. Flow rates should be limited to:

1 metre/second for copper.
3 metres/second for galvanized steel.
3 metres/second for aluminium brass.
3.5 metres/second for 90/10 cupro-nickel.
4 metres/second for 70/30 cupro-nickel.

In sea water cooling circuits, flow can be hampered and efficiency affected by build up of weed, marine life (mussels, etc.) and general deposits. External grids on hull sea water inlets also filters on the sea water system give protection. Injections of biocide should not be used where the sea water is intended for delivery to an evaporator (ref M633).

CENTRAL COOLING SYSTEMS

Where salt water corrosion is a problem it may be considered advantageous to use a closed fresh water circuit cooled by sea water in one or more large central heat exchangers. The salt water system is thereby limited to one set of pumps, valves and filters and a shorter length of piping.

The sketch (Fig. 11) shows a complete central cooling system in which all components are cooled by fresh water. The system can be divided into three main parts, (1) sea water circuit (2) high temperature circuit (3) low temperature circuit.

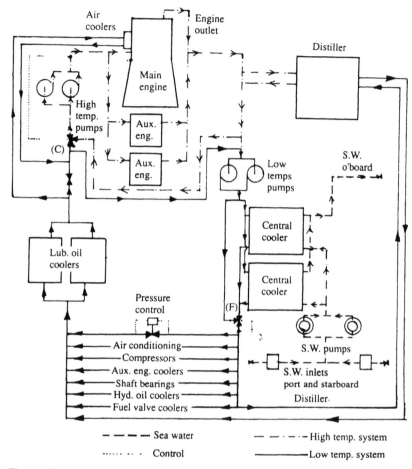

Fig. 11   Central cooling system *(Alpha Laval)*

SEA WATER CIRCUIT

The sea water pump takes water from the suctions on either side of the engine room and after passing through the cooler it is discharged straight overboard. The main and stand-by pumps would be of the double entry centrifugal type. Main circulating pumps must have direct bilge suctions, for emergency, with a diameter two thirds that of the main sea water inlet. In motor ships a direct suction on another pump of the same capacity is acceptable.

Materials for the reduced salt water system required by a central cooling arrangement will be of high quality and expensive. Savings are made by the use of cheaper metals in the protected fresh water circuit.

HIGH TEMPERATURE CIRCUIT

Cooling water for the main engine and auxiliary engines is circulated by the pumps on the left. At the outlet, the water is taken to the fresh water distiller and the heat used for evaporation of sea water. From the outlet of the fresh water distiller the water is led back to the suction of the high temperature pump through a control valve (C) which is governed by engine inlet temperature. The control valve mixes the low and high temperature streams to produce the required inlet figure—about 62°C. Outlet is about 70°C.

LOW TEMPERATURE CIRCUIT

Temperature of the water leaving the central coolers is governed by the control valve (F). Components of the system are arranged in parallel or series groups as required. The pressure control valve works on a by-pass. Temperature of the water after the cooler may be 35°C and at exit from the main engine oil coolers, it is about 45°C.

Fresh water in both the high and low temperature systems is treated chemically to prevent corrosion in the pipes and coolers.

ADVANTAGES

Provided that chemical treatment is maintained correctly corrosion will be eliminated in the fresh water system. Pipes, valves and coolers in contact with only fresh water, can be of cheaper materials. The constant temperature level of the cooling water means that control of engine coolers is easier. The number of sea water inlet valves is reduced together with the filters that require cleaning.

References

Conde, J. F. G. (1985). New Materials for the Marine and Offshore Industry. *Trans. I. Mar. E.*, vol. 97, paper 24.

Cotton, J. B. and Scholes, I. R. (1972). Titanium in Marine Engineering. *Trans. I. Mar. E.*, vol. 84, paper 16.

Shone, E. C. and Grim, G. C. (1985). 25 Years Experience with Sea Water Cooled Heat Transfer Equipment in the Shell Fleets. *Trans. I. Mar. E.*, vol. 98, paper 11.

M633 *Use of Marine Growth Inhibitors in Sea Inlet Piping.*

CHAPTER 2

Hazards in Enclosed Spaces— Tankers—Cargo Pumping

CLOSED SPACES

The fore peak is an example of a tank that can remain closed and unventilated for long periods. The atmosphere in such a tank may become deficient in oxygen due to corrosion resulting from the remains of sea water ballast. Oxygen may also be depleted by the presence in sea water of hydrogen sulphide which tends to oxidize to sulphate. Hydrogen sulphide (which is toxic) is a compound which, like ammonia, is produced by bacteria in the water, reducing sulphates and nitrogen compounds. Water containing pollutants such as hydrogen sulphide, are picked up when ballasting in estuarial waters.

While oxygen is depleted, carbon dioxide may be given off by sea water due to other chemical changes. Thus a tank, apparently safe because it has been isolated by being closed, is dangerous to enter due to lack of oxygen and sometimes the presence instead of carbon dioxide and possibly other gases. The small manhole for entry at the top of the tank will not give much assistance in ventilating the compartment. Fatal accidents have resulted from entry to fore peak tanks which have not been properly ventilated.

Normal oxygen content of air is about 21 per cent. Where a test shows that there is a lower value in an enclosed space, ventilation should be continued until the correct level is reached. The air changes necessary to improve the oxygen level will have the beneficial effect of lessening the possible presence of gas or vapour which may be harmful. A lower than normal oxygen level can cause loss of efficiency; it may cause loss of consciousness resulting in a fall with serious or even fatal injury.

In other closed compartments oxygen may be deficient due to being absorbed by chemicals or drying paint, or because it has been excluded by other gases or vapours (e.g. vapour from cargo, refrigerant, inerting gas, smoke or fumigant). Ventilation is required before entry to any ballast tank, cargo space, pumproom or closed compartment and must be maintained while work is being carried out.

OXYGEN ANALYSER

In order to measure the amount of oxygen in a sample from the atmosphere of a closed space or from flue gas etc., there must be some way of isolating it from the rest of the sample. One physical property which distinguishes oxygen from most other common gases is its paramagnetism. Faraday discovered that oxygen was paramagnetic and therefore attracted by a magnetic field. The field will also induce magnetism in the oxygen i.e. a magnetic field is intensified by the presence of oxygen and its intensity will vary with the quantity of oxygen.

Most gases are slightly diamagnetic, that is they are repelled by a magnetic field. Thus glass spheres filled with nitrogen and mounted at the ends of a bar to

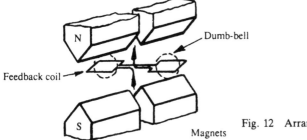

Fig. 12　Arrangement of dumb-bell in magnetic field

form a dumb-bell (Fig. 12) will tend to be pushed out from a strong symmetrical, non-uniform, magnetic field in which they are horizontally suspended.

The dumb-bell arrangement is used in the Taylor Servomex analyser. It is suspended by a platinum ribbon in the field and, being slightly diamagnetic, it takes up a position away from the most intense part of the field. The magnets and dumb-bell are housed in a chamber which has an inlet and outlet for the sample. When the surrounding gas contains oxygen, the dumb-bell spheres are pushed further out of the field due to the change produced by the paramagnetic oxygen. Torque acting on the dumb-bell is proportional to the oxygen concentration and therefore the restoring force necessary to bring the dumb-bell back to the zero position is also proportional to the oxygen concentration.

The zero position of the dumb-bell is sensed by twin photocells receiving light reflected from a mirror on the suspension. The output of the photocell is amplified and fed back to a coil wound on the dumb-bell (Fig. 13) so that the torque due to oxygen in the sample is balanced by a restoring torque generated by the feedback current. Oxygen percentage is read from the meter which measures the restoring current. This is scaled to give percentage oxygen direct. Accurate calibration is obtained by using pure nitrogen for zero and normal air for setting the span at 21 per cent oxygen.

False readings are obtained if the gas being sampled contains another paramagnetic gas. The only common gases having comparable susceptibility are nitric oxide, nitrogen dioxide and chlorine dioxide.

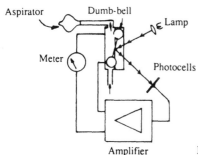

Fig. 13　Oxygen analyser *(Taylor Servomex)*

MARINE SAFETY CARD

The General Council of British Shipping has issued a safety card with precautions and a check list, to be used by personnel intending to enter a closed compartment.

OIL TANKERS

Petroleum vapours when mixed with air can be ignited provided that the mixture limits are about 1 per cent to 10 per cent of hydrocarbon vapour, with the balance, air. Below 1 per cent the mixture is too weak and above 10 per cent, too rich. These figures are termed the lower and upper flammable limits, respectively. Another condition for combustion is that the oxygen content would have to be more than 11 per cent by volume (Fig. 14).

Crude oil and the products of crude will give off vapours which are potentially dangerous if they have high or moderate volatility. Thus some petroleums will give off a lot of vapour at ordinary ambient temperature and tend to produce an over rich mixture. These are Class 'A' petroleums. This type of petroleum is dangerous because the over rich mixture is readily made flammable by dilution during loading, discharge or tank cleaning. A leak of vapour into the atmosphere is also dangerous.

Petroleums with moderate volatility give off less vapour but the amount may be within the flammable range. These are Class 'B' petroleums. The ullage space

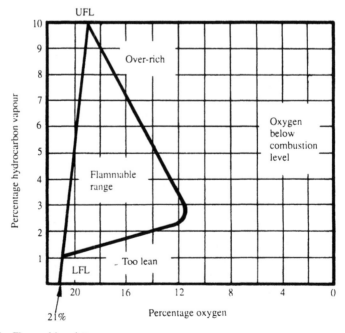

Fig. 14   Flammable mixtures

of a cargo tank containing Class B products may be filled with a flammable mixture.

Class 'C' petroleums give off little vapour unless heated to above the flash point.

EXPLOSIMETER

The atmosphere of a tank or pumproom can be tested with a combustible gas indicator which is calibrated for hydrocarbons. Frequently the scale is in terms of the lower explosive or flammable limit and marked as a percentage of the lower limit. Alternatively, the scale may be marked in parts per million (p.p.m.)

The combustible gas indicator shown diagrammatically (Fig. 15) consists of a Wheatstone bridge with current supplied from a battery. When the bridge resistances are balanced, no current flows through the meter. One resistance is a hot filament in a combustion chamber. An aspirator bulb and flexible tube are used to draw a gas sample into the chamber. The gas will burn in the presence of the red hot filament causing the temperature of the filament to rise. Rise of temperature increases the resistance of the filament and this change of resistance unbalances the bridge. Current flow registers on the meter which is scaled in percentage (L.F.L.) or p.p.m.

A lean mixture will burn in the combustion chamber, because of the filament. False readings are likely when oxygen content of the sample is low or when inert gas is present. The instrument is designed for detecting vapour in a range up to the lower flammable limit and with large percentages of gas (rich mixture) a false zero reading may also be obtained.

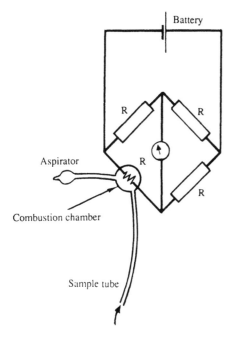

Fig. 15   Combustible gas indicator

The instrument and batteries must be tested before use and the samples are taken from as many places as possible particularly from the tank bottom. It is possible to obtain a reading for any hydrocarbon but not for the other combustible gases on an instrument which is scaled for hydrocarbons. Detection of other vapours must be by devices intended for the purpose.

The explosimeter is primarily a combustible gas detector but will also give guidance with regard to the safety of a space for entry by personnel. If a space has been ventilated to remove vapours, the remaining concentration can be measured by the explosimeter, provided that it is below the lower flammable range. Generally any needle deflection above zero is taken as indicating a toxic condition.

Crude oils contain all of the hydrocarbon products extracted in the refinery and many of the products are highly toxic. Benzene (C_6H_6) is an example and its low Threshold Limit Value (T.L.V.) of 10 p.p.m. indicates this. Sour crudes carry highly toxic hydrogen sulphide (H_2S) with a T.L.V. also of 10 p.p.m. Petrol (gasoline) has a T.L.V. of 300 p.p.m. Entry to the cargo tanks and pumprooms of a crude oil carrier exposes personnel to these risks. There are additional hazards involved with tank entry, where inert gas has been used. The inert gas adds the risk of: carbon monoxide (CO) which has a T.L.V. of 50 p.p.m.; nitrogen dioxide (NO_2) with 3 p.p.m.; nitric oxide (NO) with 25 p.p.m.; and sulphur dioxide with 2 p.p.m. Trace amounts of the hydrocarbon products which are very dangerous, and other toxic gases which may be present, require special means of detection (chemical stain tubes described below).

Threshold Limit Values are updated annually and given in references available from health and safety authorities.

THRESHOLD LIMIT VALUE

Vapour concentrations are measured in terms of parts per million (p.p.m.). A guideline, based on experience, is that personnel exposed to vapour concentrations below the T.L.V. will not be harmed but that the risk increases at concentrations above the T.L.V. Figures sometimes have to be revised because new factors come to light.

CARGO TANKS

Cargo tanks, as the result of containing Class 'A' or 'B' cargoes, may contain toxic vapours and/or flammable mixtures. Tanks which are cathodically protected and ballasted may have accumulations of hydrogen although the hydrogen will disperse with proper ventilation. Oxygen deficiency occurs due to corrosion from ballast water but as mentioned in the general section on closed compartments there are further reasons for oxygen deficiency. In a tanker, the inert gas used to produce a safe condition in tanks will reduce oxygen as will steaming out.

A gas-free tank can become dangerous again if there is sludge or scale remaining or if a pipeline containing liquid or gas is opened.

PUMPROOMS

Pumprooms are subject to leakage from pump glands and pipelines. The liquid accumulates in bilges (which should be kept clear). Volatiles will form vapour which from sour crude or benzene are toxic. Oxygen can be lacking because it has been displaced by other vapours or inert gas used in an emergency. A pumproom which has remained closed for a long period will, like a ballast tank, lose its oxygen from corrosion.

COFFERDAMS

Cofferdams and other spaces adjacent to those with cargo can become contaminated due to leakage. They may also be short of oxygen as the result of corrosion etc.

CHEMICAL TANKERS

Cargo tanks, pumprooms and other closed spaces, on chemical as on other tankers, may be deficient in oxygen or may contain flammable and/or toxic vapour. Additionally, many liquid cargoes are corrosive and toxic, others are toxic substances which are absorbed through the skin or by ingestion (swallowing). Chemical tanker problems are multiplied by the number and variety of chemicals carried and by the range of risks. Reference books are necessary and available, these containing data on the different hazards and methods of combating them.

Liquid residues in tanks and pumprooms must be considered as potentially dangerous. They may not be easily identifiable (many corrosive liquids have nothing to distinguish them from water) and the content can only be guessed at from knowledge of previous cargo. Such liquid may be corrosive, or a poison which can be absorbed through the skin. If volatile, the vapour may be toxic or flammable.

Contamination by gases and vapours can be checked provided that presence of a particular substance is suspected.

CHEMICAL STAIN TUBES

Testing for contaminants in spaces where the variety precludes the provision of special detecting instruments, is made possible with the use of tubes packed with chemical granules that change colour on contact with a particular gas or vapour. The glass stain tubes are sealed to protect the detecting chemicals, the ends being broken off before use. There is a different chemical tube for each toxic substance.

The pump or aspirator (Fig. 16) for taking the test has a long sample tube on its suction side and the stain tube is fitted in the discharge after first purging to clear air and fill the tube with a sample from the space. The result, obtained using Dräger instruments, can be read directly from the tube after the prescribed number of pumping strokes. Chemicals have a two-year shelf life.

GENERAL PRECAUTIONS

Any closed space requires ventilation before entry and the ventilation must be maintained during the time that work is being carried out. If there is reason to suspect lack of oxygen or the presence of toxic vapours then ventilation or gas freeing is started some time before entry and the atmosphere is checked before going in.

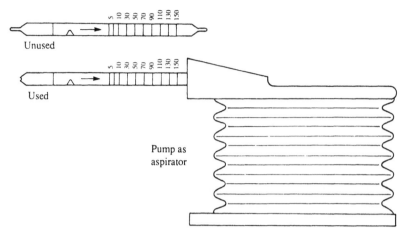

Fig. 16   Dräger type multi-gas detector

Liquids in bilges or in the bottoms of tanks should be drained as far as possible. The liquid may itself be dangerous, or dangerous vapour may be produced if it is volatile. Scales and sludges, particularly when heated, tend to give off vapours. These should be removed before entry. Valves and pipelines should not be opened as contamination can occur from liquid or vapour in the pipe.

Before working on oil or chemical cargo pumps or pipelines, they should be washed through. After pumping certain chemicals, it may be necessary to use a solvent to wash the pump. As an extra precaution, protective clothing and breathing apparatus may be necessary. Work on pumps should only be carried out when tanks are in a safe condition. Hydraulic or steam lines to pump motors must be closed securely and the power system shut down.

Any portable lights used must be gas tight and safe.

When a closed space has to be entered, there must be a second person in attendance at the entrance or, if it is a routine, other personnel should be made aware that entry is intended. The duty of the watcher at the entrance or anybody else involved, is to go for assistance in the event of trouble.

When the situation requires the wearing of breathing apparatus and lifeline, the set must be thoroughly checked and signals arranged with those in attendance.

The location of rescue equipment and the method of using it must be generally known.

GAS FREE CERTIFICATE

When work has to be carried out in port a certificate may be required stating that a space is gas free or that it is safe for hot work (welding, burning etc.). The certificate is obtained from an authorized chemist after tests to prove that any gas present is below the lower flammable limit and below the T.L.V. in quantity. As well as being gas free, the compartment must also be free from oil or other residues and from scale or sludge.

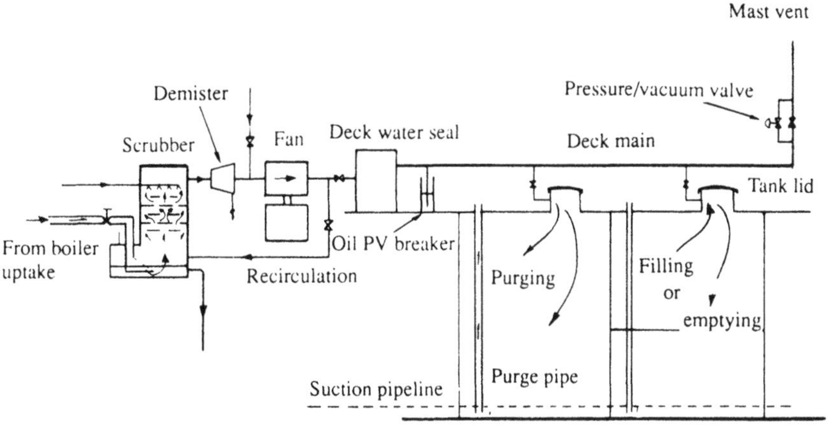

Mast vent

Demister

Pressure/vacuum valve

Scrubber   Fan   Deck water seal   Deck main

Tank lid

From boiler   Oil PV breaker
uptake   Recirculation   Purging   Filling
or
emptying

Suction pipeline   Purge pipe

Fig. 17   Inert gas system

INERT GAS SYSTEM

The presence of flammable vapours in the cargo tanks of oil tankers has led to
the development of inert gas systems. Washed flue gas from either the main
boiler or from a special gas generator is used for the purpose.

With good combustion, flue gas will contain about 3 per cent to 5 per cent
oxygen which is below the figure required for combustion (see Fig. 14). The gas is
drawn from the boiler uptake and passed through a scrubber, where it is cooled
by a sea water spray which also washes out corrosive sulphur oxides and solids.
The gas is then pumped into the tanks by the fan. At start up the gas is
recirculated through the scrubber or a start up vent is used. To prevent
hydrocarbon gases from the tanks from passing back through the system, the gas
is pumped through the deck water seal, which acts as a non-return valve (Fig.
18). A second safety device is the pressure/vacuum breaker.

Combustion in the boiler is controlled to give the minimum oxygen and the
tank atmospheres are checked by oxygen analyser when inerting or venting. The
system fan can be used to ventilate the tanks for entry by personnel.

When loading or ballasting, the incoming liquid displaces the gas through the
mast vent. During unloading, the gas is pumped in as cargo (or ballast) is
pumped out. The system is used as required when loading or tank cleaning to
bring the tanks to a safe condition.

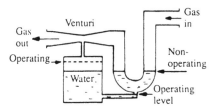

Fig. 18   Howden type deck water seal

23

Inert tanks have low oxygen content and before entry by personnel, must be ventilated and tested.

PRESSURE/VACUUM VALVES

Moderate pressures of 0.24 bar (3½ lb/in^2) acting on large surfaces in liquid cargo tanks are sufficient to cause damage and rupture. The pressure on each unit of area multiplied by the total area gives a large loading on the underside of the top of a tank or other surface, which may then buckle or the metal plate may be torn. Similarly, pressure drop within a tank can cause damage due to greater atmospheric pressure on the outside. Pressure/vacuum valves (Fig. 19) in the ventilation system will prevent either over or under pressure. They are set usually so that tank pressure of about 0.14 bar (2 lb/in^2) will lift the main valve (the smaller valve will lift with it) and release excess pressure. The vapour passes to atmosphere through a gauze flame trap. Drop in tank pressure compared with that of the outside atmosphere will make the small valve open downwards to equalise internal pressure with that outside.

Pressure vacuum valves can relieve moderate changes in tank pressure due to variations of temperature and vapour quantity. A drop towards vacuum conditions as the result of the condensation of steam will also be handled by the valve. Rapid pressure rise due to an explosion would not be relieved.

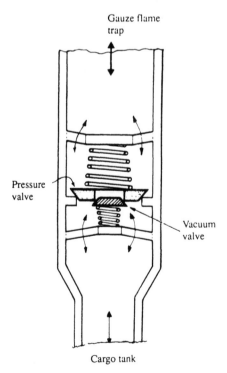

Gauze flame trap

Pressure valve

Vacuum valve

Cargo tank

Fig. 19   Pressure/vacuum valve

The fast rate at which a tank is filled while loading produces a very rapid expulsion of the previous contents (vapour and inert gas). The pressure vacuum valve is not designed as a filling vent and neither should the tank hatch be left open. The latter method of venting can cause an accumulation of flammable vapours at deck level. Tanks should be vented while filling, through masthead vents (see Fig. 17) or through special high velocity vents.

HIGH VELOCITY VENTS

Tank vapours can be released and sent clear of the decks during loading through large, high velocity vents. The type shown (Fig. 20) has a moving orifice, held down by a counterweight to seal around the bottom of a fixed cone. Pressure build up in the tank as filling proceeds causes the moving orifice to lift. The small gap between orifice lip and the fixed cone gives high velocity to the emitted vapour. It is directed upwards with an estimated velocity of 30 metres per second. Air drawn in by the ejector effect dilutes the plume.

The conical flame screen fixed to the moving orifice to give protection against flame travel will, like the moving parts, require periodic cleaning to remove gummy deposit. The cover is closed (as shown) when the vessel is on passage.

A simpler design of vent (Fig. 21), has two weighted flaps which are pushed open by pressure build up to achieve a similar nozzle effect.

OIL TANKER CARGO PUMPING

Centrifugal cargo pumps with a double entry impeller have largely replaced reciprocating pumps in oil tankers. These pumps are cheaper, have no suction or delivery valves, pistons, rings, etc and therefore require less maintenance. The

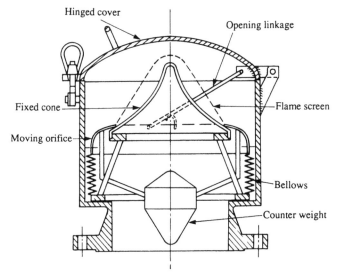

Fig. 20   High velocity vent *(IOTTA)*

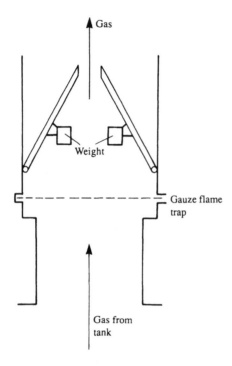

Fig. 21 Simple high velocity vent

compact centrifugal pump can be mounted horizontally or vertically in the pump room with a turbine, or in some ships electric motor, drive from the engine room. The drive shaft passes through the engine room bulkhead via a gas-tight seal.

Rate of pumping is high ($2600m^3$/hr) until a low level is reached, when loss of head and impeded flow through frames and limber holes makes slowdown in the rate of pumping necessary if use of a small stripping pump is to be avoided. Systems such as the Worthington–Simpson 'Vac-Strip' enable a faster general rate of discharge to be maintained while reducing the rate of discharge at lower tank levels to allow for draining. The basic parts of such a system are shown in Fig. 22.

Suction from the cargo tank is taken through a separator tank to the pump inlet and discharge from the pump is through a butterfly valve to the deck main. When cargo tank level drops and flow is less than the rate of pumping, liquid level in the separator tank will also reduce and this will be registered by the level monitoring device. The latter will automatically start the vacuum pump and cause the opening of a diaphragm valve to allow passage of vapour to the vacuum pump from the separator tank. General accumulation of vapour in the suction tank will cause the same result. The vacuum pump will prime the system by removing air or vapour. Rise of liquid in the separator tank will cause the vacuum pump and vapour valve to be closed down.

Continuing drop in liquid level due to slow draining necessitates a slowdown in the pumping rate and this is achieved by throttling of the main pump butterfly

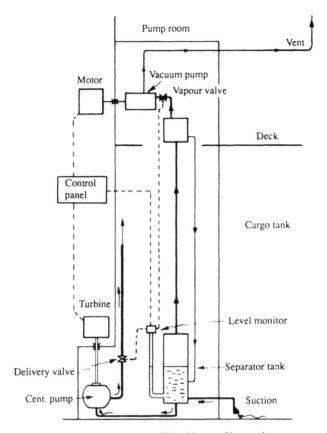

Fig. 22 'Vac-strip' type pumping system (Worthington-Simpson)

discharge valve. Valve closure is controlled by the level monitoring device. The butterfly valve can also be hand operated. Throttling is not harmful to the centrifugal pump in the short term.

The primer/vacuum pump driven by an electric motor in the engine room is of the water ring type as described in the previous chapter. Its shaft has a gas seal.

CHEMICAL TANKER PUMPING AND TANK DRAINAGE

Pump rooms in chemical tankers are dangerous because of the risk of leakage from pump glands of toxic/flammable vapour and corrosive or otherwise harmful liquids. The practice of positioning submersible or deepwell pumps within cargo tanks eliminates pump room dangers, also the expense of extra suction pipework and the risk of mixing cargoes with resulting contamination. Deepwell pumps are described in the section on liquefied gas pumping. To make them

27

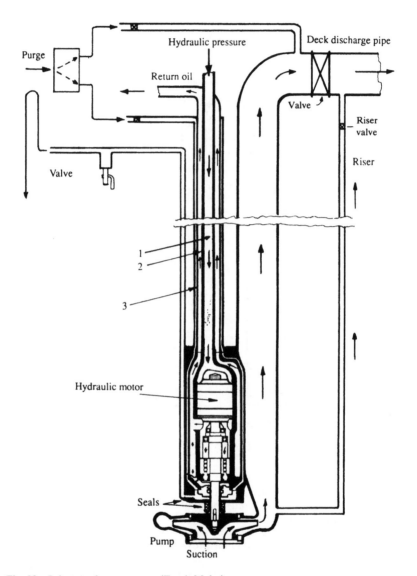

Fig. 23 Submerged cargo pump (Frank Mohn)

suitable for chemical pumping there will be a different gland arrangement and shaft bearings of Teflon.

An example of a submerged pump is shown in Fig. 23. Three concentric tubes make up: the high pressure oil supply pipe (1) to the hydraulic motor, the return pipe (2), and a protective outer cofferdam (3). As stainless steel, although expensive, will withstand the corrosive effects of most chemicals it may be used for the pipework, pump and casing. Obviously, where a chemical tanker is to be engaged on a particular trade it may be possible to use cheaper materials. Working pressure for the hydraulic circuit is up to about 170 bar and return pressure about 3 bar.

The impeller suction is positioned close to the bottom of the suction well for good tank drainage but when pumping is completed the vertical discharge pipe will be left full of liquid. Stopping the pump would allow the liquid to fall back into the tank and clearing of the tank of cargo or of water used in tank cleaning would be a constant problem. Thus purging connections are fitted to clear the discharge pipe (and the cofferdam if there is leakage). Discharge pipe purging is effected by closing the deck discharge valve as the tank clears of liquid, then with the pump left running to prevent cargo fallback opening the purge connection shown. The compressed air or inert gas at 7 bar will clear the vertical discharge pipe by pressurising it from the top and forcing liquid cargo up through the small riser to the deck main.

The cofferdam is also pressurised before the pump is stopped, to check for leakage. This safety cofferdam around the hydraulic pipes is connected to the drainage chamber at the bottom of the pump. Seals above and below the chamber exclude ingress of low pressure hydraulic oil and liquid cargo from the tank, respectively. The bottom seal is subject only to pressure from the head of cargo in the tank, not to pump pressure.

LIQUEFIED GAS CARGO PUMPING

The low temperature of liquefied gas prohibits the use of submerged hydraulically driven pumps. Thus deepwell pumps (Fig. 24) are fitted or submerged electrically driven pumps.

The long shaft of the deepwell pump runs in carbon bearings, the shaft being protected in way of the bearings by stainless steel sleeves. Positioning of the shaft within the discharge pipe allows the liquid cargo to lubricate and cool the bearings. There is a risk of overheated bearings if the pump is run without flow of cargo and therefore a pressure cut-out or thermal switch may be arranged.

Liquefied gas is carried at its boiling temperature, which ensures that the ullage space above the liquid is filled with cargo vapour, air being excluded. The vapour although flammable is safe within the tank because of the lack of any oxygen (air) to burn with. External leakage is dangerous.

At the end of cargo discharge a residue of about 2% liquid is left to maintain a totally vapour atmosphere in the tank. This remainder keeps the tank filled with vapour for safety and saves the problems of total drainage and inerting. Only when the tank has to be emptied for repair, etc is it drained and purged with inert gas and then air. The small amount of cargo left also allows tank temperature to be kept at the carrying level so the expense of cooling down and the probability of damage to the tank structure caused by expansion and contraction is avoided.

The weight of the pump shaft and impellers is considerable and one or more carrier bearings are fitted. Lift of the shaft due to ship movement or pump action also requires a downward-acting thrust bearing.

The number of stages in the multi-stage pump shown is dictated by the discharge head required. As an alternative to the multi-stage unit and its high-power drive through a long line shaft, a single stage low-pressure pump can be

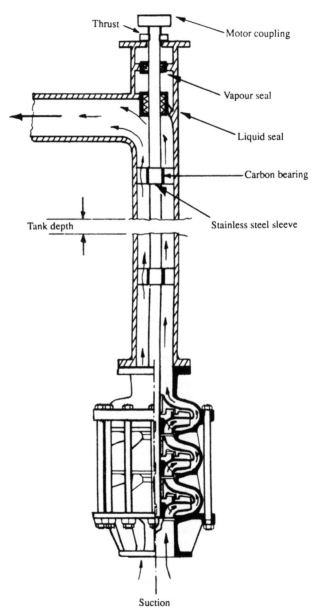

Fig. 24   Deepwell pump

fitted in the tank to lift the liquid to deck level where there is a booster pump to transfer the cargo ashore.

The inducer frequently fitted to centrifugal liquefied gas pumps is an archimedian screw attached to the drive shaft just below the impeller at the pump suction. It is useful for inducing flow into the pump, particularly from volatile liquids.

Deepwell pumps are driven by hydraulic motors or by flameproof electric motors situated at deck level. Duplication of pumps in tanks is the safeguard against breakdown of deepwell pumps in liquid gas carriers.

CRUDE OIL WASHING OF TANKS (COW)

Crude oil washing of cargo tanks is carried out while the vessel is discharging, with high-pressure jets of crude oil. For the process, a portion of the cargo is diverted through fixed piping to permanently positioned tank cleaning nozzles. Suspended nozzles can be controlled to give a spray pattern on the top area of the tank sides and then progressively further down as surfaces are uncovered during the discharge. Bottom washing is timed to coincide with the tank emptying so that oil below heating coil level will not solidify in cold weather. Effective washing can be carried out at the recommended heating temperature for discharge and even at temperatures as low as 5°C above the pour point. The waxy and asphaltic residues are readily dissolved in the crude oil of which they were previously a part and better results are obtained than with water washing. The oil residues are pumped ashore with the cargo. An inert gas system must be in use during tank cleaning.

Crude oil washing is necessary on a routine basis for preventing excessive accumulation of sludge. Tanks are washed at least every four months. Unless sludge is regularly removed drainage will be slow. If it is likely that ballast may have to be carried in cargo tanks (additional ballast to that in segregated ballast tanks, for example because of bad weather) then suitable tanks are crude oil washed and water rinsed. Water ballasted into a dirty cargo tank in emergency would be discharged in compliance with the anti-pollution regulations and a suitable entry would be made in the Oil Record Book (see Chapter 9). Crude oil washing must be completed before the vessel leaves port: a completely different routine from that of water washing which, where used, is carried out between ports.

Clean Ballast Tanks (CBT) One of the aims of the Regulations for the Prevention of Pollution by Oil (Marpol) which came into force in October 1983 is to reduce the practice of using cargo tanks for sea water ballast alternatively with cargo. As a temporary measure, older tankers have been permitted to dedicate certain cargo tanks to be used for ballast only. Cargo piping may be used for the introduction and discharge of the ballast, however.

Segregated Ballast Tanks (SBT) New large tankers are required to be built with an adequate number of ballast (only) tanks, so that under normal circumstances water will not have to be carried in cargo tanks. Ballast is handled by means other than the cargo discharge pipe system.

The new procedures have greatly reduced the potential for pollution from discharge of dirty ballast (formerly taken in cargo tanks) and tank cleaning.

References

ICS (1988). *International Safety Guide for Oil Tankers and Terminals*, 3rd edn. International Chamber of Shipping.

ICS (1971). *Tanker Safety Guide (Chemicals)*. International Chamber of Shipping.

ICS (1978). *Tanker Safety Guide (Liquefied Gas)*. International Chamber of Shipping.

Day, C. F., Platt, E. H. W., Telfer, I. E., and Tetreau, R. P. (1972). The Development and Operation of an Inert Gas System for Oil Tankers. *Trans. I. Mar. E.*, vol. 84.

McGuire, G., and White, B. (1986). *Liquefied Gas Handling Principles on Ships and in Terminals*. SIGTTO.

CHAPTER 3

Fire Protection

FIRE MAIN

Water is the chief fire fighting medium on a ship and the fire main is the basic installation for fighting fires. The system shown (Fig. 25) has two independently powered pumps which are also used for general service and ballast. These pumps supply two engine room hydrants and the deck main through the isolating valve. The latter is required to prevent loss of water through damaged pipework in the engine room if, to maintain the deck supply, the emergency fire pump has to be used. The emergency fire pump is shown as being situated in the tunnel, with a supply to the deck main through the tunnel escape and also to both hydrants in the shaft tunnel by the engine room watertight door. The deck main has a drain at the lowest position so that the pipe can be emptied (particularly of fresh water) in cold weather. If this is not done, the pipe can be damaged by the water freezing but more important, it will be blocked by the ice and not usable.

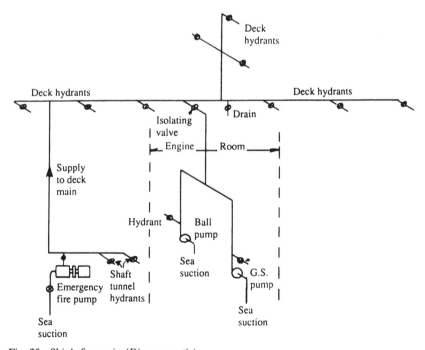

Fig. 25   Ship's fire main *(Diagrammatic)*

PUMPS

Two independently powered pumps must be provided in all cargo ships of 1000 tons gross and over and in passenger ships of less than 4000 tons gross. Larger passenger vessels must have three such pumps. The pumps are fitted with non-return valves if they are of the centrifugal type, to prevent loss of water back through open valves when not running. A relief valve is necessary in the system if the pumps are capable of raising the pipeline pressure so that it is greater than the design figure. Minimum line pressures and capacities are governed by the regulations. While fire pumps may be used for other duties such as ballast, bilge or general service they should not normally be used for pumping oil. Changeover arrangements are fitted if a pump can be used for oily bilges etc.

EMERGENCY FIRE PUMP

Normally, cargo vessels are provided with emergency fire pumps because a fire in the engine room could put all of the other pumps out of action. Such a pump is indicated in Fig. 25, and is located away from the engine room in the tunnel, steering gear or in the forward part of the ship. The suction lift of any pump is limited and for this reason emergency pumps are usually at a maximum of 6 metres from the water level at light draught, or installed below water level. On

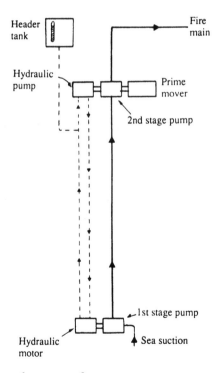

Fig. 26   Arrangement of emergency fire pump

large vessels a special two stage pump arrangement may be used (Fig. 26). The first stage below the waterline is driven by a hydraulic motor. The second stage and the hydraulic power unit are driven from the prime mover which can be positioned at more than the normal distance from the waterline.

An emergency pump has an independent diesel drive or some alternative such as an electric motor powered from the emergency generator, or an air operated pump with its own air supply.

PIPELINES ETC.

Where steel pipes are used, they are galvanized after bending and welding. Diameter is between 50 mm and 178 mm depending on the size and type of ship. Engine room hydrants must have hoses and nozzles for jet and fog or dual purpose nozzles (Fig. 27).

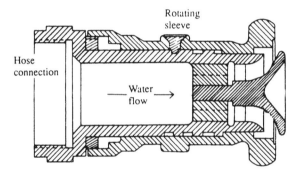

Fig. 27   Dual purpose nozzle

INTERNATIONAL SHORE CONNECTION

The shore connection (Fig. 28) is a standard sized flange with nuts, bolts and washers and a coupling for the ship's fittings. The dimensions are shown. The fitting and joint must be suitable for a working pressure of 10.5 bar. Four bolts are required of 16 mm diameter and 50 mm length, also eight washers.

HOSES AND NOZZLES

Fire hoses must be of approved material. They are positioned adjacent to hydrants together with suitable nozzles. Dual purpose nozzles (Fig. 27) can be adjusted by rotation of the sleeve to produce a jet or spray. These are an alternative to having one nozzle for a jet and another for a spray or fog to be used for oil fires.

Foamite branchpipes (Fig. 29) similar to those used in deck installations for tankers are fitted for use with the hydrants in some machinery spaces, car decks etc. These are available in various sizes for operation at a range of pressures and outputs. The branchpipe is connected through a hose to the hydrant and the water flow produces a venturi effect which draws up Foamite liquid through the pick-up tube, from a container. The action also draws in air. Mixing of the three

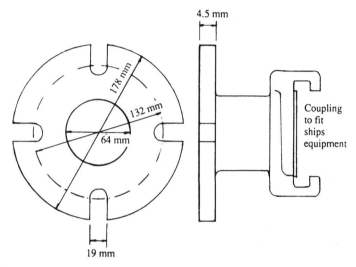

Coupling
to fit
ships
equipment

Fig. 28   International shore connection

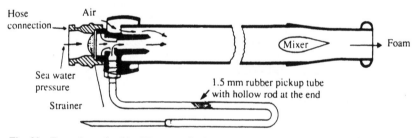

Fig. 29   Foam branch pipe *(Merryweather)*

components in the tube causes formation of a jet of foam. Initially, only water issues from the branchpipe and the nozzle is directed away from the fire until foam appears. When the foam compound is exhausted, water will again appear at the nozzle. Foam continuity is achieved by dropping the pick-up tube in a bucket and keeping the bucket topped up with foam liquid.

MACHINERY SPACE FIXED FIRE EXTINGUISHING INSTALLATIONS

Engine room spaces are protected by fixed fire extinguishing installations. As with the emergency stops and shut-offs, the equipment must be operable from a position outside of the space.

SMOTHERING GAS INSTALLATION

The carbon dioxide system shown (Fig. 30) consists of bottles of CO_2 with a gang release arrangement and a pipe to the engine room distribution nozzles via a master valve.

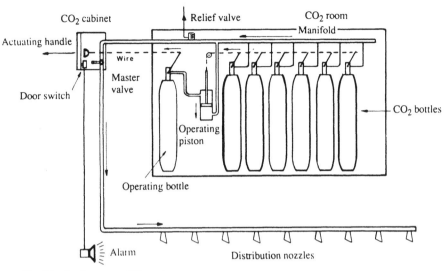

Fig. 30   Engine room CO_2 system

The CO_2 system is used if a fire is severe enough to force evacuation of the engine room. An alarm is sounded by an alarm button as the CO_2 cabinet is opened and in some ships there is also a stop for the engine room fans incorporated (Fig. 31). Before releasing the CO_2 personnel must be counted and the engine room must be in a shut down condition with all openings and vent

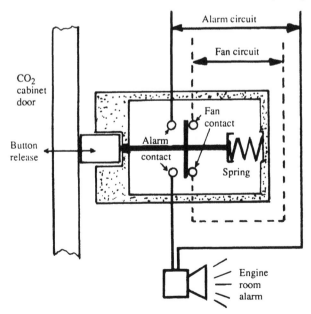

Fig. 31   Alarm and fan switch operated by CO_2 cabinet door

flaps closed. It is a requirement that 85 per cent of the required quantity of gas is released into the space within two minutes of operating the actuating handle.

In the system shown, the actuating handle opens an operating bottle of CO_2 and the gas from this pushes down the piston to release the other bottles. To avoid sticking, all the handles must be in good alignment. The bottle valves may be of the quick-release type (Fig. 32) where the combined seal/bursting disc is pierced by a cutter. The latter is hollow for passage of liquid CO_2 to the discharge pipe.

Bottle pressure is normally about 52 bar (750 lb/in^2) but this varies with temperature. Bottles should not be stored where the temperaure is likely to exceed 55°C. The seal/bursting discs are designed to rupture spontaneously at pressures of 177 bar produced by a temperature of about 63°C. The master valve prevents CO_2 released in this way from reaching the engine room and it is dispersed safely by the relief on the manifold.

Rapid injection of CO_2 is necessary to combat an engine room fire which has attained such magnitude that the space has to be vacated. This is the reason for the rule that 85 per cent of the gas must be released within two minutes. The quantity of gas carried (a) must be sufficient to give a free gas volume equal to 40

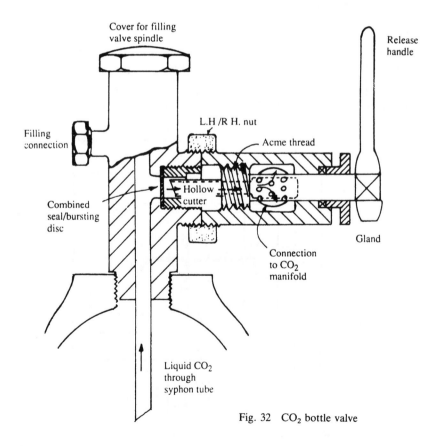

Fig. 32 CO_2 bottle valve

per cent of the volume of the space except where the horizontal casing area is less than 40 per cent of the general area of the space, or (b) must give a free gas volume equal to 35 per cent of the entire space; whichever is greater. The free air volume of air receivers may have to be taken into consideration.

The closing of all engine room openings and vent flaps will prevent entry of air to the space. All fans and pumps for fuel, can be shut down remotely as can valves on fuel pipes from fuel service and storage tanks.

CO_2 bottles are of solid drawn steel, hydraulically tested to 228 bar. The contents are checked by weighing or by means of a radioactive level indicator. Recharging is necessary if there is a 10 per cent weight loss.

Pipework is of solid drawn mild steel, galvanized for protection against corrosion. The syphon tube in the bottle ensures that liquid is discharged from the bottles. Without the syphon tube the CO_2 would evaporate from the surface and, taking latent heat, would cause the remaining CO_2 in the bottle to freeze.

BULK CO_2

An alternative to the conventional CO_2 storage in individual bottles for large installations is the storage of the liquid in bulk. Refrigeration units are required for this form of containment.

Storage pressure for bulk CO_2 is 21 bar and the temperature maintained in the bottles is $-17°C$. A suitable steel for this temperature would probably contain 3½ per cent nickel. Nickel used for low temperature steels reduces the coefficient of expansion and resultant thermal stress. The pressure vessels are constructed to Lloyds Class 1 standard.

Two refrigeration units, each capable of maintaining the required temperature, are provided. Failure of one unit causes automatic starting of the other. Failure is indicated by an alarm.

Vessels are safeguarded against abnormal pressure increase by relief valves set to 24.5 bar. The discharge from these valves is piped away from the CO_2 storage space to a safe area. Relief valves set to the higher figure of 27 bar are also fitted and arranged for discharge into the space to extinguish a local fire causing the pressure rise. The discharge line has a relief valve set to 35 bar.

Continuous contents monitoring is provided by a remote reading electrical gauge. A stand-by indicator is required in addition and provided by a vertical, external, uninsulated pipe, which can be filled with liquid CO_2 to the vessel level, by opening one valve. Liquid level is shown by frosting or by a radioactive device as used for CO_2 bottles.

Isolating valves, are of the bellows sealed globe type. The main CO_2 discharge line is sensed for pressure so that release of gas is indicated by an alarm.

HALON SYSTEM

Halon is an alternative gas for use as a fire fighting medium in the engine room fixed installation. It is a halogenated compound made by replacement of hydrogen in methane or ethane by one of the halogens. Fluorine, chlorine and bromine are halogens. The compound is Bromo-Trifluoro-Methane (BTM), $CBrF_3$.

As a fire fighting agent, the gas operates not by smothering as does CO_2 but chemically by acting as a negative catalyst to inhibit combustion by breaking the combustion reaction. CO_2 by being introduced in large quantities produces a relative drop in oxygen content which requires evacuation of the space before use. Smaller quantities of halon 1301 are required because it has a different action. The quantity is calculated as about 5 per cent of the volume of the space to be protected. This amount is not harmful to personnel for up to five minutes, provided that breakdown of the gas has not occurred to any great extent. The gas starts to breakdown at temperatures over 510°C. The products are toxic (e.g. hydrogen fluoride and hydrogen bromide) but being irritants they give warning of their presence.

The harmful products are increased by the intensity of the fire. To reduce this effect, detection of the fire and discharge of the gas must be rapid. Speed of discharge of the gas is also important, because of its high rate of dispersal. A maximum discharge time of 20 seconds is called for, compared with up to two minutes for CO_2.

The gas is colourless, odourless, of high density and has a low boiling point, thus it can be stored as a liquid like CO_2. Storage pressure is low and to ensure rapid discharge the liquid is further pressurized with nitrogen or CO_2.

An arrangement is made for automatic stopping of vent fans and closure of dampers in conjunction with the sounding of the space alarm at opening of the door of the release cabinet. The high dispersal rate of the gas makes sealing of the space essential.

The halon is stored in a compartment away from the protected space. The operating cabinet is also remote. The gas is effective for the same types of fire as CO_2. It is not suitable for certain fires involving metals or metal hydrides nor is it normally installed in chemical tankers for cargo tank protection.

SYSTEM OPERATION

The halon release arrangement shown (Fig. 33) consists of the storage container, two sets of CO_2 operating cylinders and a manual release cabinet. The halon is stored at a pressure of 14 bar in the container which has a pressure relief, filling valve and level-indicator.

Release procedure is much the same as for CO_2. When the cabinet is opened, the alarm operates, fans stop and dampers will close. With all entrances closed, the handles (1) and (2) are operated in succession. Handle (2) can only be moved when released by the blocking mechanism.

The contents of the CO_2 bottles opened by handle (1) pressurize the pipe line between the halon container and the master valve causing the bursting disc to rupture and allowing the halon to flow as far as the master valve. The pressure build up in this line acts on the blocking device to permit operation of handle (2). The latter opens the master valve to the engine room distribution pipe and also opens the CO_2 bottles (2). CO_2 from these ruptures the bursting disc at the top of the storage container and then assists in expelling the halon.

The discharge must be completed in 20 seconds but the alarm sounds as the release cabinet door is opened. Personnel must evacuate the space when warned. The 5 per cent concentration gives risk and is treated with the same caution as CO_2.

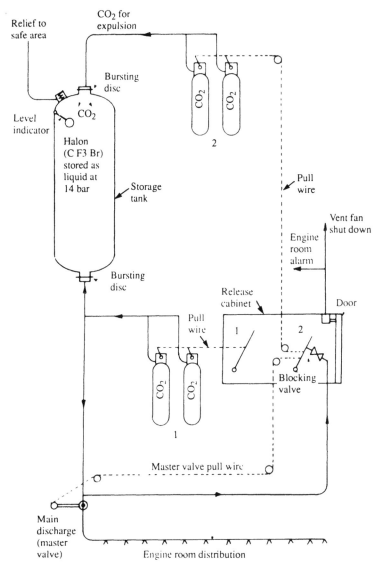

Fig. 33 Halon release system

FIXED PRESSURE WATER SPRAYING SYSTEM

This system is similar to the sprinkler used in accommodation areas but the spray heads are not operated automatically. The section control valves (Fig. 34) are opened by hand to supply water to the heads in one section or to a hose. Fresh water is used for the initial filling and the system is brought to working

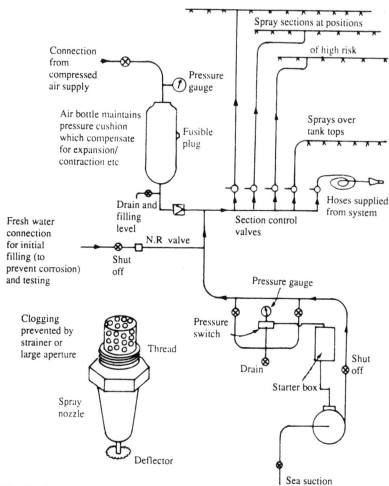

Fig. 34 Engine room fixed pressure water spray (multi-spray)

pressure by means of the compressed air connection. The air bottle provides a cushion and prevents cut-in of the pump due to slight leakage of the water. The pump is automatically operated by pressure drop in the system when the control valve to one section is opened.

The pump must have either an independent drive or an electrical supply from the emergency generator. It must be able to maintain working pressure when supplying all the sections simultaneously in one compartment. It is placed outside the compartment.

Spray nozzles are designed to give the correct droplet size for fires in flammable liquids such as fuels and lubricating oils, when working at the correct pressure. They are located so as to give adequate water distribution over the tank top and all fire risk areas.

Water spray is a potentially good fire fighting medium because
(a) it produces a large quantity of steam which has a smothering action
(b) in producing the steam, a large amount of heat is required (latent heat) and this gives a cooling effect
(c) the spray will protect personnel in the compartment
(d) water is easily available.

Corrosion of the system is reduced by keeping it normally filled with fresh water. After operation, the pipework is drained of salt water and refilled with fresh after washing through. Damage to electrical equipment etc. by salt water is reduced by washing with hot fresh water before drying out.

FOAM SYSTEM

A number of large tankers have been fitted with low expansion or heavy foam installations for protection both on deck and in the machinery spaces.

The system shown (Fig. 35) is designed to deliver a correctly proportioned amount of foam compound into the water supply to the deck main. The foam is drawn from one of two large tanks fitted with vent arrangements by the foam pump. The pump itself has a relief valve but foam compound excess to requirements is discharged back to the tank via the diaphragm valve, which is controlled by two sensing lines. As demand varies due to the number of outlets, the diaphragm valve will deliver the correct amount of foam compound into the water main for any set of conditions.

Foam monitors are fitted on deck and supplied through the deck main. There are additional portable nozzles, also supplied through the deck main.

Foam drencher nozzles in the machinery space are situated above areas of high risk. The control valves are located in the foam compartment which is outside of the machinery space. It must be possible to operate the system from outside the protected space. A second water supply is available from the emergency fire pump normally situated in the fore part of a tanker. On a large vessel this pump may be of the type described in the firemain section i.e. a two stage pump with the first stage below the waterline, which acts as a booster for the second stage, fitted at deck level. The foam plant must be capable of providing foam for fire fighting in the machinery space when water is supplied by the emergency fire pump (pressure may be less from the emergency pump than from the main fire pump).

High expansion foam is generated by blowing air through a mesh which has been wetted by a solution of foam concentrate in water. It has been used for hold protection on some container vessels and has been tested for engine room fire fighting.

The mesh is corrugated and its hole size governs the expansion ratio of the foam which is limited to 1000:1 by IMO rules. The limit is required because the foam is composed largely of air and easily breaks down when in contact with a fire. However, in the 1000:1 foam, the original 1 volume of liquid evaporates and produces enough steam to reduce the percentage oxygen in the steam/air mixture to about 7.5 per cent. This amount is below the level normally required for combustion. Heavier foams can be produced.

The foam concentrate is metered or mixed with the water to give a 1.5 per

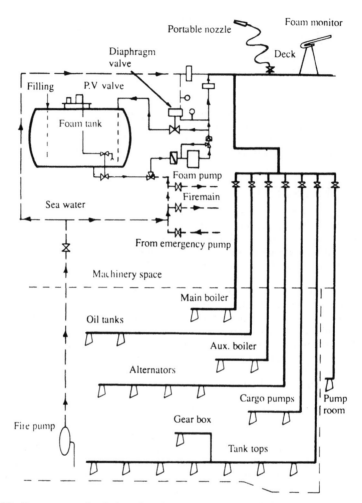

Fig. 35 Foam system for deck and engine room *(Merryweather)*

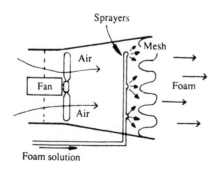

Fig. 36 High expansion foam generator

cent solution of concentrate in water, and sprayed onto the screen. Air is blown through by an electrically driven fan (other drives have been used). Delivery ducts are necessary to carry the foam to the fire area but normal ventilation trunkings may be acceptable.

Generation of foam must be rapid and sufficient to fill the largest space to be protected at the rate of 1 metre/minute (depth).

SPRINKLER SYSTEM ACCOMMODATION

Accommodation and service spaces of passenger vessels are protected against fire by an automatic sprinkler fire alarm/detection system of the type shown (Fig. 37).

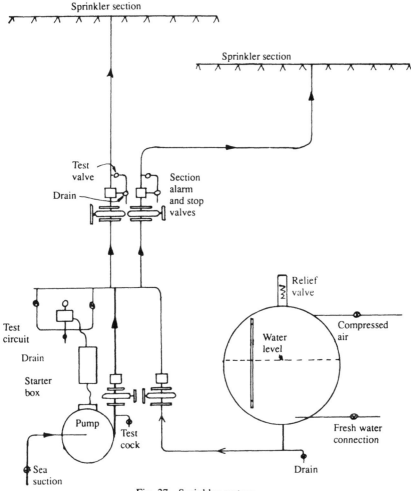

Fig. 37 Sprinkler system

The installation is kept ready for use by being pressurized from the compressed air loading on the water in the pressure tank. Local temperature rise will operate an individual sprinkler head and the resulting pressure drop in the system will, through the pressure switch, start the pump. Water supply is then maintained by the pump. Sprinkler heads are fitted throughout the accommodation but the number of heads in any section is limited to about 150 (IMO regulations allow up to 200). Pipes are of steel galvanized for corrosion protection and the system is filled initially with fresh water. If part of the circuit becomes filled with salt water as the result of operation, it must be subsequently drained and again filled with fresh water. A test circuit is fitted so that the pressure switch can be isolated and, by draining, be caused to start the pump (the pump test cock is opened to give a water flow).

PUMP

The pump sea suction is kept permanently open and the pump is for use only with the sprinkler system. It must be connected to main and emergency power supplies and there must be a connection through a screw down non-return valve (locked) from the fire main such that backflow from the sprinkler to the fire main is prevented.

TANK

Contents level is indicated by a gauge glass and this is specified as equivalent to one minute's discharge of the pump. Tank volume is at least twice that of the water specified. Initially the flow of water through a sprinkler system relies on the pressure exerted by the compressed air. It must be arranged that during expulsion of the standing charge of fresh water from the tank, the air pressure remains sufficient to overcome the head to the highest sprinkler and to provide enough working pressure for the sprinkler. Thus a compressor connected and having automatic starting is provided. Supply pressure must be higher than that in the tank so that air can be replenished under pressure. The tank is also provided with a fresh water supply, drain and relief valve. The pipe from the tank to the system has a non-return valve to prevent entry of sea water to the tank.

SECTION ALARM VALVE

The sprinklers are grouped in sections with a limited number of heads. Normally a section is confined to one fire zone or area. There is a stop valve for each section either locked open or fitted with a tell-tale alarm to prevent unauthorized closure. A special non-return valve (Fig. 38) is the means for operating the alarm and visual indicator which is positioned on the bridge/fire control centre. The indicator is a panel showing a section of the fire zones with an alarm light for each. Thus when a sprinkler head is set off, the approximate location is displayed.

The alarm is a pressure switch and water from the system reaches it when the non-return valve is lifted by water flowing to any sprinkler head. Normally the non-return valve covers the annular channel in the seat. The drip orifice prevents build up of water pressure in the alarm pipe due to leakage. Section pressure is shown by the gauge at each stop valve and a drilled hole through the

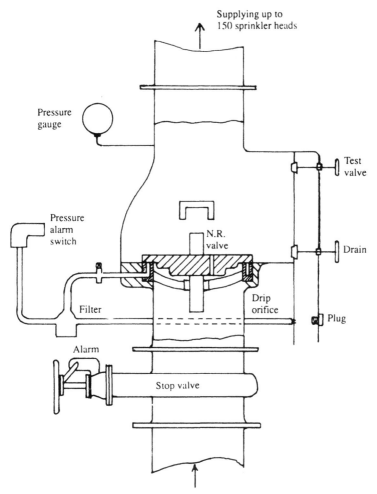

Supplying up to
150 sprinkler heads

Pressure
gauge

Test
valve

Pressure
alarm
switch

N.R.
valve

Drain

Filter

Drip
orifice

Plug

Alarm

Stop valve

Fig. 38 Section alarm for sprinkler

valve accommodates expansion due to temperature change. The test valve gives a discharge of water equivalent to that of one sprinkler and is used to test the section alarm.

Any parts of the system which might be subjected to freezing, must be protected. Anti-freeze can be added to fresh water for this purpose. Some vessels which trade in low temperature areas, have dry pipe sections installed.

DRY PIPE SECTION ALARM

The dry pipe (Fig. 39) extends upwards from the section valve which also acts as the link between the sprinkler system water pressure and the dry pipe pressurized with air.

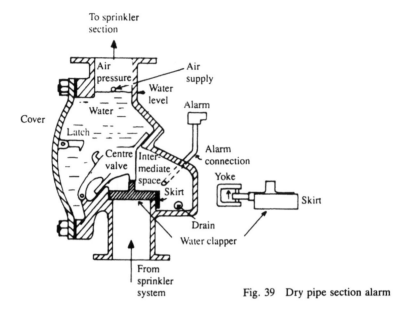

Fig. 39 Dry pipe section alarm

Water pressure is contained by the water clapper which is held on its seat by the centre valve. The space above the centre valve is filled to the level with water and the pipe above that is filled with air under pressure. The centre valve is made watertight by a joint and the intermediate space is dry.

When operation of a sprinkler head releases the pressure in the dry pipe, the centre valve is pushed up by the force of water under the clapper. The clapper lifts and rotates on the yoke, being swung to one side by the effect of water flow on the skirt. The water floods up through the dry pipe causing the centre valve to lock open and in filling the intermediate chamber, pressurizes and operates the alarm.

Pressure gauges for air and water are required. The section valve opens when air pressure drops to ⅙th that of the water pressure.

The cover has to be removed to reset the valve.

SPRINKLER HEAD

Sprinklers must give a minimum cover of 5 litres/m^2 per minute over the protected area.

The head (Fig. 40) is closed by a valve which is held in place by a quartzoid bulb. Excessive heat causes the bulb to shatter by expansion of the liquid it contains. Heads are designed to act at a particular temperature and the bulbs are colour coded (red: 68°C, yellow: 79°C). Sprinklers in the accommodation are normally these types. Higher temperature sprinkler heads are fitted if necessary.

Spares must be carried and the temperature code is marked on the deflector.

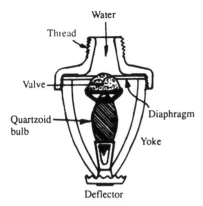

Water

Thread

Valve

Quartzoid
bulb

Diaphragm

Yoke

Deflector

Fig. 40   Sprinkler head

FIRE ALARM ONLY

Thermal or smoke detector heads can be fitted in the accommodation as an alternative to the sprinkler system. The heads are similar to those used for machinery spaces.

CHAPTER 4

Fuel—Handling and Treatment— Self Cleaning Purifier— Automatic Combustion System for Auxiliary Boiler

CRUDE OIL REFINING

Crude oil is broken down to give various types of fuel and other substances, primarily by heating it so that vapours are boiled off and then condensed at different temperatures, and separately collecting the constituents or 'fractions' from the distillation process. Crude oil contains gaseous fuels, petrols, paraffins, gas oils, distillate diesel fuels and lubricating oils which can be collected from the fractionating tower, where they are condensed out at the different levels maintained at appropriate temperatures. The crude is heated in a furnace, shown with the tower on the left of Fig. 41.

The boiling process leaves behind a residue which is very dense as the result of having lost the lighter parts. This high-density remainder has the same hydrocar-

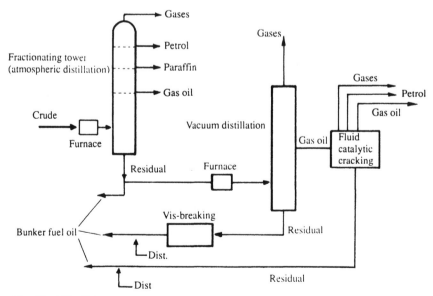

Fig. 41 Oil refining processes

bon make-up as the lighter fractions and is potentially a fuel. Unfortunately, the initial refining process not only concentrates the liquid but also the impurities.

Vacuum distillation removes more of the lighter fractions, to leave an even heavier residue. As can be seen from the flow chart (Fig. 41) the refinery may have additional conversion equipment. Vis-breaking or thermal cracking is one process using heat and pressure to split heavy molecules into lighter components leaving heavy residue. Catalytic cracking is another, that uses a powdered silica-alumina based catalyst with more moderate temperatures and pressures to obtain lighter fractions, with however an increasingly heavy residue. In the latter process catalyst powder is continuously circulated through the reactor, then to a regenerator where carbon picked up during the conversion reaction is burnt off. Unfortunately, some of the catalyst powder remains in the residue which may be used for blending bunker fuel oils. The very abrasive silica-alumina catalytic fines have caused severe engine wear when not detected and removed by slow purification in the ship's fuel treatment system.

QUALITY ASSESSMENT (FUEL TESTING)

Bunkers are classified as Gas Oil, Light and Marine Diesel Oil, Intermediate Fuel Oil and Marine or Bunker (C) Fuel Oil. The delivery note specifies the type(s) of fuel, amount(s), viscosity, specific gravity, flash point and water content. Trouble frequently results from inferior fuels and there is insufficient information to give warning. Fuel grading schemes and detailed delivery notes have been proposed. Thus the Classification Societies and others provide fuel testing services (about ten days for a result), and on-board testing equipment is available. A representative sample is needed to give an accurate test result and this is difficult to obtain unless a properly situated test cock is fitted in the bunker manifold where flow is turbulent. The sample is taken after flushing of the test cock. Because of the variation in heavy fuel, small quantities are tapped into the test container over the period of bunkering, to give a representative sample.

A full analysis can be given by the shore laboratory. On-board tests are limited to those which give reliable results and kits for specific gravity, viscosity, pour point, water content and compatibility are on the market. Flash point is found with a Pensky–Martin closed cup apparatus as carried by passenger vessels.

FUEL DATA

Density or **Specific Gravity (S.G.)** is given on the bunker delivery note and can be found for fuel at a particular temperature with a test kit hydrometer. Required for bunker calculations, it is also needed for selection of the purifier gravity disc. Separation of water from fuels with a density higher than water is not possible in conventional centrifuges, although solids will separate.

Viscosity is specified when ordering fuel, Redwood Viscosity in seconds being replaced by Kinematic Viscosity in centistokes. The on-board test uses the principle of a metal ball being allowed to fall through a tube of fuel sample at 80°C and being timed over a set distance after it reaches its terminal velocity. A calculation gives the kinematic viscosity. Fuel viscosity is reduced by heating, (1) to make it usable in the engine, (2) to improve separation and (3) to ensure ease of pumping.

Pour Point is checked on the ship by cooling a pre-heated oil in a testtube while tilting the tube at every 3°C drop in temperature to see whether it is still free-flowing. The pour point is 3°C above the no-flow temperature, and fuel in tanks must be 5°C above this to prevent solidification. Low heat transfer coefficient makes a solidified fuel impossible to re-liquefy in the tank.

Water Content is found with a test kit which is the same as that used for measuring water content of lubricating oils. The oil sample is mixed with a reagent in a closed container and any water in the oil reacts with the chemical reagent to produce a vapour. Pressure rise due to the vapour generation is registered on a pressure gauge, calibrated to show water percentage. Alternatively, the vapour displaces liquid and the quantity of this is used to show water content. The water may be fresh or sea water. The latter is a source of sodium, which with vanadium produces harmful ash after combustion of the fuel. Water together with solids dissolved in it is normally removed in the centrifuge.

Compatibility of a residual fuel allows it to be blended with other fuels to give a stable mix. Where there is incompatibility, the mixing of fuels results in the precipitation of heavy sludge which blocks the fuel system. On-board blending and even mixing of different fuels in bunker tanks can cause the problem of sludge, if there is incompatibility. The property is assessed by making a sample mix with the residual fuel and another fuel, in equal amounts, and depositing a drop on photochromatic paper. After an hour the pattern left by the dried drop is compared with spots on a reference sheet. An unsatisfactory blend is typified by a dark central deposit with an outer less dark ring. Economies are possible if poor-quality fuel can be sufficiently improved by blending with distillate to allow its use in generator engines, etc but there is the risk of sludge formation. Fuels blended ashore and supplied as bunkers are sometimes found to be unstable and subsequent sludge formation or layering may make them unusable. The compatibility test is not reliable.

FUEL HANDLING AND TREATMENT

Heating coils in bunker tanks and particularly in double bottom tanks have always been necessary for heating to make the fuel pumpable. Some modern poor-quality fuels have a pour point sufficiently high that they start to solidify at normal temperatures, and because of low conductivity they cannot be re-liquefied by heating. With these fuels the temperature in storage must be 5°C above pour point. High delivery temperature is a sign of high pour point fuel.

The risk of incompatibility between fuels, and the possible precipitation of sludge if bunkers are taken in tanks that contain remains of another fuel, has increased. Bunkers should therefore be taken in empty tanks and kept segregated. Formation of sludge can also result from sea water contamination or temperature changes. Bacterial infestation is well known as a problem with lubricating oil. It has also caused sludge and other problems with all types of fuel.

Filter blockage may give notice of sludge in the service tank if it carries over from the purifier. Sludge will choke the fuel system and stop the engine.

Settling tanks and centrifuges are intended to remove water and solid contaminants; and although small amounts of water take a long time to settle out in the settling tank, large quantities from serious water contamination will be drained when the tank is sludged and the indication will be given that a serious problem exists. Thus settling tanks are not made redundant by centrifuges.

Centrifuges remove water with the impurities dissolved in it, and the heavier solids in suspension, at a rapid rate. Impurities dissolved in the fuel are not removed. There are fuels with high density over 991 kg/cm^3 at 15°C from which water separation in the normal way is impossible.

Sodium is a contaminant which may be present due to sea water in the fuel. Being soluble in water, it will be removed with it in the purifier, (Sodium in the cylinder burns to form with vanadium an ash very harmful to engine exhaust valves, etc.)

Vanadium and **sulphur** are dissolved in the fuel and not removed by the purifier. The effect of sulphur (which burns to sulphur oxides and forms corrosive sulphuric acid) can be neutralised by alkali additives in diesel engine cylinder lubricant. Condensation of strong acid is diminished by keeping liner temperatures high. Sticking of sodium and vanadium salts on exhaust valves is much reduced by keeping valve working temperature below about 530°C.

Silicon and **aluminium** catalytic fines are removed in the purifier or clarifier, but slow throughput is essential for optimum results. The self-driven purifier delivery pump has been replaced in some installations by one with independent drive to give a slower supply of fuel to the purifier.

SELF CLEANING PURIFIER

Fuel oil purifiers are fitted with a sliding bowl bottom which is raised or lowered by a water operating system for self cleaning. Lubricating oil purifiers are sometimes self cleaning. Manual cleaning may be preferred so that the solids can be examined and also because use may be intermittent and the extra expense not justified. Fig. 42 shows one operating system for a self cleaning purifier.

While oil is passing through the purifier, the sliding bowl bottom is held up in the position shown by the operating water beneath it. The sliding bottom seals the bowl by being pressed against the sealing ring in the rim of the cover. Solids from the oil are thrown outwards by centrifugal force and collect against the bowl periphery. At intervals dictated by either a timer or choice, the oil feed is turned off and the bowl opened to discharge the solids. There are a number of discharge ports around the bowl. At the end of the discharge, the bowl is closed and after the liquid seal has been re-established, the oil feed is continued.

During normal running, the pressure exerted by the water under the sliding bottom, is sufficient to keep it closed against the pressure from the liquid in the bowl. The operating water tank maintains a constant head of water to the paring disc which acts like a pump opposing this head, provided that the radius of the liquid annulus remains constant. If evaporation or leakage causes a slight water loss, the reverse pumping effect of the paring disc is reduced and water from the operating tank brings the quantity of water in the paring chamber back to the correct radius. The operating slide prevents loss of water from beneath the sliding bowl, by closing the drain holes.

To discharge the solids, first the oil feed is closed and then the solenoid valve is opened. This allows water from the high pressure line to flow into the paring chamber. The water enters from a point nearer the centre than the normal

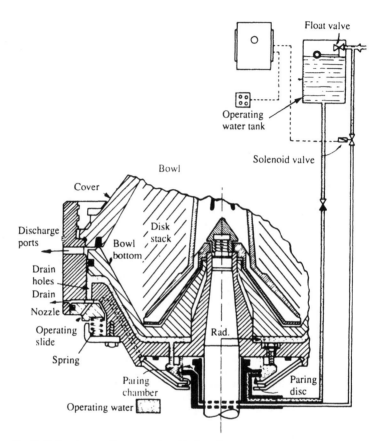

Fig. 42   Self cleaning purifier operation (de Laval)

radius at which the operating water is maintained. This extra water (indicated by the arrow) fills the paring chamber until it runs over the lip and via the drilling in the bowl body, into the 'opening chamber' immediately above the operating slide.

Water in the operating chamber, builds up a pressure due to centrifugal force (despite small loss through the drain nozzle) which pushes the operating slide down against the springs beneath it. As soon as the operating slide begins to move downwards, the drain holes open and the high pressure operating water under the sliding bowl escapes rapidly. Pressure exerted by liquid in the bowl forces the bottom down and solids are discharged through the ports.

When all of the operating water has drained from the underside of the sliding bottom and discharge of solids is complete, then with the solenoid valve closed the operating slide is moved back up by the springs to close the drain holes. To raise the sliding bottom, the chamber under it must be filled with operating water. The filling is completed quickly by a short opening of the solenoid valve. When the chamber is filled and pressurized the paring chamber will start to fill.

At this point the solenoid is closed to prevent overflow and a second opening. The radius of the liquid annulus is then maintained by the operating tank and paring disc arrangement.

CENTRIFUGE ARRANGEMENTS

An arrangement of purifier and clarifier was fitted on many ships in the earlier days of burning residual fuel in slow-running diesels. Fuel quality was better then and where the clarifier was found to contain very little residue or none its use was discontinued. A single purifier gave adequate results until the advent of silicon and aluminium fines in fuels from refineries with the catalytic cracking process. Abrasion from the fine powder impurity made up from the silicon and aluminium fines has variously ruined fuel pumps and injection equipment, also piston rings and liners. Research at this time showed that a combination of two purifiers in parallel (to give very slow throughput), or the old arrangement of purifier and clarifier in series, would remove the majority of fines and make the fuel usable without the risk of serious engine damage. Latterly the installation of two purifiers in parallel with a third centrifuge as clarifier in series has been recommended.

The latest Alfa–Laval design of centrifuge is a self-sludging machine which has a flow control disc that makes it virtually a clarifier. There are no longer gravity discs to be changed to make the machine suitable for fuels of different specific gravity/density. Water and sludge accumulate in the outer part of the bowl as the result of centrifugal effect and as the interface moves inwards, but before reaching the disc stack the water flows through to reach a water sensing transducer that causes, via micro-processor circuitry, the bowl to self-sludge. The system is said to be capable of handling fuels of greater than 991 kg/m^3 density. A figure for successful water separation from fuels with densities as high as 1010 kg/m^3 at 15°C is given.

FUEL BLENDER

Conventionally, the main engine uses cheaper heavy fuel oil and generators lighter, more expensive, distillate fuel. The addition of a small amount of diesel oil to heavy fuel considerably reduces its viscosity, and if heating is used to further bring the viscosity down then the blend can be used in generators with resultant savings.

The in-line blender shown (Fig. 43) takes fuels from heavy oil and light diesel tanks, then mixes and supplies the mix direct to the auxiliary diesels. Returning fuel is accepted back in the blender circulating line. It is not directed back to a tank where there would be the danger of the two fuels settling out.

Fuel is circulated around the closed loop of the system by the circulating pump against the back-pressure of the p.s. (pressure sustaining) valve. Thus there is supply pressure for the engine before the valve, and a low enough pressure after it, to allow returning oil back into the loop. Sufficient light diesel is injected into the loop by the metering pump for light load running. As increased load demands more fuel this is drawn in from the heavy oil tank by a drop in loop pressure on the suction side of the circulating pump. Heavy fuel thus makes up the extra, made necessary by load increase. At full load the ratio may be 30% diesel with 70% heavy fuel.

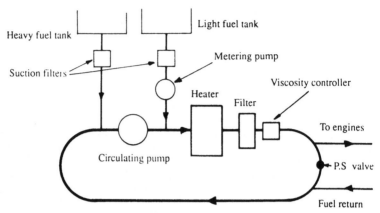

Fig. 43   Sea star in-line blender

A viscotherm monitors viscosity and controls it through the heater. The hot filter removes particles down to 5 micron size and there are other filters on the tank suctions. Constant circulation and remixing of the blend and the returning fuel prevents separation.

HOMOGENISER

The homogeniser provides another solution to the problem of water in high-density fuels. It can be used to emulsify a small percentage for injection into the engine with the fuel. This is in contradiction to the normal aim of removing water, which in the free state can cause gassing of fuel pumps and other problems. However, experiments in fuel economy have led to the installation of homogenisers on some ships to deal with a deliberate mixture of up to 10% water in fuel. The homogeniser is fitted in the pipeline between service tank and engine so that the fuel is used immediately. It is suggested that a high-density fuel could be emulsified and burnt in the engine. A homogeniser could not be used in place of a purifier for diesel fuel as it does not remove abrasives like aluminium and silicon, or ash-forming sodium which damages exhaust valves.

The three disc stacks in the rotating carrier of the Vickers type homogeniser (Fig. 44) are turned at about 1200 rev/min. Their freedom to move radially outwards means that the centrifugal effect throws them hard against the lining tyre of the homogeniser casing. Pressure and the rotating contact breaks down sludges and water trapped between discs and tyre, and with the general spinning action aids mixing.

Auxiliary Boilers

PACKAGE BOILER COMBUSTION SYSTEM

A simple automatic combustion system is required for auxiliary boilers of the package type. Many small boilers both ashore and those used at sea, are fitted

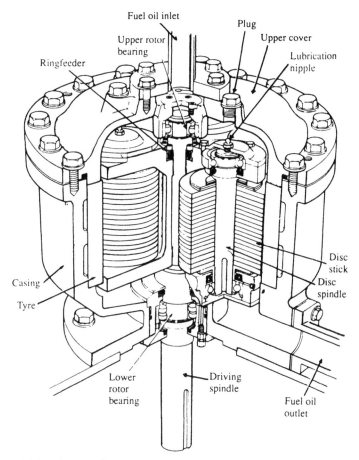

Fig. 44   Vickers homogenizer

with the arrangement shown (Fig. 45—The burner is drawn oversize to show detail).

The burner has a spring loaded piston valve which closes off the passage to the atomising nozzle when fuel is supplied to the burner at low pressure. If the fuel pressure is increased the piston valve will be opened so that fuel passes through the atomiser. The atomiser can be supplied with fuel at different pressures by the system.

The solenoid valves are two-way, in that the fuel entering can be delivered through either of two outlets. The spill valves are spring loaded. When either one is in circuit, it provides the only return path for the fuel to the suction side of the fuel pressure pump. The pressure in the circuit will be forced, therefore, to build up to the setting of the spill valve.

A gear pump with a relief arrangement to prevent excessive pressure, is used

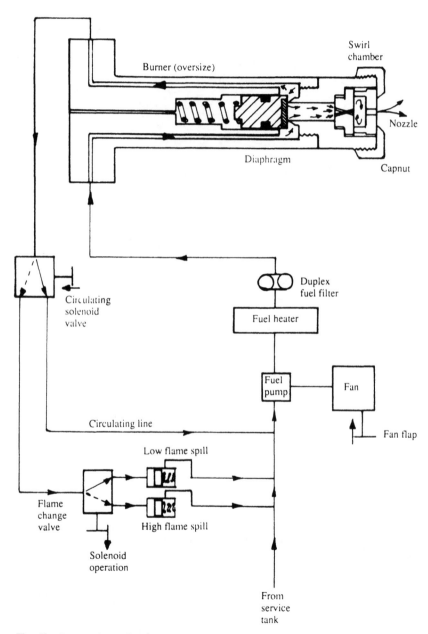

Fig. 45 Automatic combustion system

to supply fuel to the burner. Fuel pressure is varied by the operation of the system and may range up to 40 bar.

Combustion air is supplied by a constant speed fan, and a damper arrangement is used to change the setting.

OPERATION

Electrical circuits are arranged so that when the boiler is switched on (assuming water level is correct etc.) the system will (1) heat up and circulate the fuel (2) purge the combustion space of unburnt gas (3) ignite the flame and, by controlling it, maintain the steam pressure.

When the boiler is started current is supplied first to the fuel heater. The electric heating elements are thermostatically controlled and when oil in the heater reaches the required atomising temperature, another thermostat switches in the fan and oil circulating pump. Air from the fan purges the combustion spaces for a set time, which must be sufficient to clear the gases completely, otherwise an air/gas explosive mixture may be formed. The oil circulates from the pump and heater through the system via the oil circulating valve. This ensures that the oil in the burner is hot and thin enough to atomise.

When the oil circulating solenoid is operated, the fuel no longer returns to the suction side of the pump but is delivered to the low flame spill through the oil change valve. With the ignition arc 'on', oil pressure builds up sufficiently to open the piston valve in the burner. The atomised fuel is ignited and once the flame is established, control of the oil change valve and fan damper depends on steam pressure. With low steam pressure, the oil control valve is actuated to deliver the fuel to the high flame spill. When steam pressure rises, the fuel is switched back to the low flame spill. The fan damper is operated at the same time to adjust the air delivery to the high or low flame requirement. The solenoids or pulling motor for the operation of the high/low flame devices are controlled by a pressure switch acted on by boiler steam pressure.

SAFETY DEVICES

Package boilers have the normal safety devices fitted to boilers and also special arrangements for unattended operation.

FLAME FAILURE

The flame is monitored by photo-cells. If the flame goes out abnormally or ignition fails, the photo-cell shuts down the combustion system and causes the alarm to sound. Sometimes trouble with combustion will have the same effect if the protective glass over the cell becomes smoke blackened.

LOW WATER LEVEL

Water level is maintained by a feed pump controlled by a float-operated on/off switch. The float chamber is external to the boiler and connected by pipes to the steam and water spaces. There is a drain at the bottom of the float chamber. A similar float switch is fitted to activate an alarm and shut-down in the event of low water level (and high water level on some installations). Because float chambers and gauge glasses are at the water level, they can become choked by solids which tend to form a surface scum on the water. Gauge glasses must be regularly checked by blowing the steam and water cocks through the drain. When float chambers are tested, caution is needed to avoid damage to the float. Frequent scumming will remove the solids.

STEAM PRESSURE

The boiler pressure will stay within the working range if the pressure switch is set to match output. If a fault develops or steam demand drops, then high steam pressure will cause the burner to cut out.

AIR SUPPLY

Incorrect air quantity due to a fault with the damper would cause poor combustion. Air delivery should be monitored.

FUEL TEMPERATURE

Many package boilers burn a light fuel and heating is not required. Where a heater is in use, deviation from the correct temperature will cause the burner to be shut off.

TESTING CONTROLS

The automatic combustion system is checked when the boiler is started. The flame failure photo-cell may be masked or some means such as starting the boiler with the circulating solenoid cut out, may be used to ensure flame failure shut down. Cut outs for protection against low water level, excess steam pressure, loss of air and change of fuel temperature are also checked. Tests necessary vary with different boilers. At shut down the air purge should operate, the fan being set to continue running for a limited time.

FEED PUMP

Before starting the boiler, feed tank level and pump suction filters are checked. It may be necessary to test the feed water. When the boiler is in use the feed pump is checked.

CHAPTER 5

Refrigeration—
Air Conditioning—Heating

VAPOUR COMPRESSION CYCLE

The basic components of any refrigeration system working on the vapour compression cycle are the compressor, condenser, expansion valve and evaporator. The refrigerant with which the circuit is charged is normally R 12 or R 22 (i.e. Freon 12 or Freon 22).

Compressor

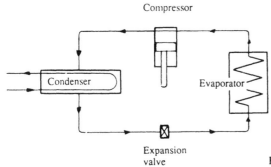

Expansion
valve   Fig. 46   Simple fridge system

COMPRESSOR

The temperature at which a fluid boils or condenses is known as the saturation temperature, and varies with pressure. The compressor in a refrigeration system, in raising the pressure of the vapourized refrigerant, causes its saturation temperature to rise so that it is higher than that of the sea water or air cooling the condenser. The compressor also promotes circulation of the refrigerant by pumping it around the system.

CONDENSER

In the condenser the refrigerant is liquefied and subcooled to below the saturation temperature by the circulating sea water or air. Latent heat, originally from the evaporator, is thus transferred to the cooling medium. The liquid refrigerant, still at the pressure produced by the compressor, passes to the expansion valve.

EXPANSION VALVE

The expansion valve is the regulator through which the refrigerant flows from the high pressure side of the system to the low pressure side. The pressure drop causes the saturation temperature of the refrigerant to fall so that it will boil at the low temperature of the evaporator. In fact as the liquid passes through the expansion valve the pressure drop causes its saturation temperature to fall below its actual temperature. The result is that some of the liquid boils off at the expansion valve taking latent heat from the remainder and causing its temperature to drop.

The expansion valve throttles the liquid refrigerant and maintains the pressure difference between the condenser and evaporator while supplying refrigerant to the evaporator at the correct rate. It is thermostatically controlled.

EVAPORATOR

The refrigerant entering the evaporator at a temperature lower than the secondary coolant (air or brine) receives latent heat and evaporates. Later this heat is given up in the condenser when the refrigerant is again liquefied.

In a small refrigerator the evaporator cools without forced circulation of a secondary coolant. In larger installations the evaporator cools air or brine. These are circulated in turn as secondary refrigerants.

REFRIGERANT

There is no perfect refrigerant for all operating conditions. Refrigerants in general use are selected as being nearest to the ideal.

REFRIGERANT 12

For cargo installations R 12 replaced carbon dioxide and is in turn being replaced by R 22. It is a halogenated hydrocarbon derived from methane (CH_4) with the hydrogen having been displaced by chlorine and fluoride. The resulting compound is dichlorodifluoromethane (CCl_2F_2) also known as Freon 12. The trade name of Freons was given to a number of similar compounds.

Refrigerant 12 is considered to be non-toxic except in high concentrations producing oxygen deficiency. However, it decomposes in contact with flame to give products which are pungent and poisonous (chlorine Cl_2 and phosgene $COCl_2$). The gas escaping under pressure will cause skin damage on contact. It is odourless, non-irritant and not considered flammable or an explosive.

Working pressures and temperatures are moderate and the high critical temperature (112°C) is well above the working range.

REFRIGERANT 22

The popularity of R 22 as a refrigerant for cargo installations has increased in recent years at the expense of R 12. It is more suitable for a lower temperature range than R 12 because the pressure on the evaporator side of the system is higher than atmospheric at low temperatures (thus reducing the risk of drawing air into the system). Its performance is better, approaching that of ammonia.

The chemical and other properties are similar to R 12 except that it is not miscible with oil over the full temperature range. The compound is Chlorodifluoromethane ($CHClF_2$).

REFRIGERANT 11

For air conditioning installations R 11 (monofluorotrichloromethane (CCl_3F) has been found suitable.

REFRIGERANT 502

This refrigerant is composed of 48.8 per cent of R 22 and 51.2 per cent of R 115 (C_2ClF_5). It is particularly suited for use with hermetic compressors.

CARBON DIOXIDE

When carbon dioxide (CO_2) is used as a refrigerant the working pressures are high, being about 70 bar at the compressor discharge and 20 bar at the compressor suction. The machinery and system must therefore be of substantial construction. The critical temperature is low (31°C) and this causes problems in areas with high sea water temperature. It also has a low coefficient of performance.

The gas is not explosive or flammable but a leak is potentially dangerous because it can displace air and asphyxiate. The liquid is stored in steel bottles at high pressure, ideally in a cool space. Temperature rise will cause pressure rise in the bottles which is relieved by the rupturing of a safety disc and release of the gas.

AMMONIA (R 717)

Thermodynamically, ammonia is a good refrigerant but it is explosive, poisonous and an irritant. The explosive mixture is 16 to 25 per cent in air. It is corrosive to copper and its alloys so that ferrous materials are used for components in a system employing ammonia.

Ammonia is a reactive compound. It is highly soluble in water with which it forms ammonium hydroxide a weak base. About 1300 volumes of ammonia can be dissolved in 1 volume of water at low temperature however it is easily expelled by boiling. This action makes the vapour absorption refrigerator possible. The high solubility in water also means that a wet cloth held to the face will give some protection against an ammonia leak in an emergency although a breathing apparatus would be worn in such a case, normally.

Because of the hazards, ammonia is used mainly ashore and on fishing vessels.

AUTOMATIC FREON SYSTEM

The circuit shown in Fig. 47 contains the basic compressor, condenser, expansion valve evaporator and also the controls for automatic operation.

The compressor is started and stopped by the L.P. (low pressure) controller in response to the pressure in the compressor suction. There is also an H.P. (high pressure) cut-out with a hand re-set which operates to shut down the compressor in the event of high discharge pressure. The compressor will supply a number of cold compartments through thermostatically controlled solenoids. Thus as each room temperature is brought down, its solenoid will close off the liquid refrigerant to that space. When all compartment solenoids are shut, the pressure

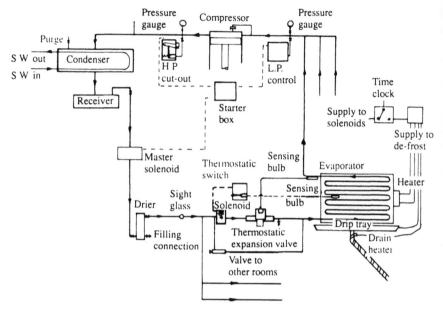

Fig. 47   Automatic Freon system

drop in the compressor suction will cause the compressor to be stopped through the L.P. controller. Subsequent rise of compartment temperature will cause the solenoids to be re-opened by the room thermostats. Pressure rise in the compressor suction acts through the L.P. controller to restart the compressor.

Each cold compartment has a thermostatic expansion valve, as the regulator through which the correct amount of refrigerant is passed.

On large systems a master solenoid may be fitted. If the compressor stops due to a fault, the master solenoid will close to prevent flooding by liquid refrigerant and possible compressor damage.

The sketch is for a three compartment system but only shows the details for one. Each room has a solenoid, regulator and evaporator. Air blown through the evaporator coils acts as the secondary refrigerant. Regular defrosting by means of electric heating elements keeps the evaporator free from ice. The time switch de-energizes the solenoids to shut down the system and supplies the power to the heaters instead.

SYSTEM COMPONENTS

The pressure gauge on the compressor discharge shows the gas pressure and also has marked on it the relative condensing temperature. This is not the actual temperature of the gas which is higher and shown by the thermometer. The pressure gauge should show a pressure with an equivalent temperature about 7° or 8°C above the sea water inlet.

HIGH PRESSURE CUT-OUT

When the compressor is running the liquid refrigerant is pumped around the circuit in the direction of the arrows. In the event of overpressure on the condenser side of the compressor, the H.P. cut-out will cause the compressor to shut down. The device is re-set by hand. There are a number of faults which can cause high discharge pressure, included in the section on faults.

The bellows in the cut-out (Fig. 48) is connected by a small bore pipe between the compressor discharge and the condenser. The bellows tend to be expanded by the pressure and this movement is opposed by the spring. The adjustment screw is used to set the spring pressure.

During normal system operation, the switch arm is held up by the switch arm catch and holds the electrical contact in place. Excessive pressure expands the bellows and moves the switch arm catch around its pivot. The upper end slips to the right of the step and releases the switch arm so breaking the electrical contact and causing the compressor to cut-out. The machine cannot be restarted until the trouble has been remedied and the switch re-set by hand.

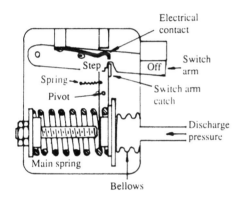

Fig. 48   H P. cut out

THERMOSTATIC EXPANSION VALVE

This is the regulator through which the refrigerant passes from the high pressure side of the system to the low pressure side. The pressure drop causes the evaporating temperature of the refrigerant to fall below that of the evaporator. Thus the refrigerant can be boiled off by an evaporator temperature of $-18°C$ because the pressure drop brings the evaporating temperature of the refrigerant to say $-24°C$.

The liquid refrigerant leaves the condenser with a temperature just above that of the sea water inlet, say 15°C. As it passes through the expansion valve the evaporating temperature decreases to $-24°C$ and some of the liquid boils off taking its latent heat from the remainder of the liquid and reducing its temperature to below that of the evaporator.

The aperture in the expansion valve is controlled by pressure variation on the top of a bellows. This is effective through the push pins (Fig. 49) and tends to open the valve against the spring. Spring pressure is set during manufacture of

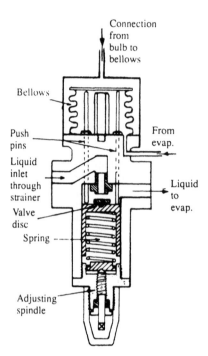

Connection
from
bulb to
bellows

Bellows

Push
pins

From
evap.

Liquid
inlet
through
strainer

Liquid
to
evap.

Valve
disc

Spring

Adjusting
spindle

Fig. 49   Thermostatic expansion valve

the valve and should not be adjusted. The pressure on the bellows is from a closed system of heat sensitive fluid in a bulb and capillary connected to the top of the bellows casing. The bulb (Fig. 50) is fastened to the outside of the evaporator outlet so that temperature changes in the gas leaving the evaporator are sensed by expansion or contraction of the fluid. Ideally the gas should leave with 6° or 7°C of superheat. This ensures that the refrigerant is being used efficiently and that no liquid reaches the compressor. A starved condition in the evaporator will result in a greater superheat which through expansion of the liquid in the bulb and capillary, will cause the valve to open further and increase the flow of refrigerant. A flooded evaporator will result in lower superheat and the valve will decrease the flow of refrigerant by closing in as pressure on the top of the bellows reduces.

Saturation temperature is related to pressure but the addition of superheat to a gas or vapour occurs after the latent heat transaction has ended. The actual pressure at the end of an evaporator coil is produced inside the bellows by the equalizing line and this is in effect more than balanced by the pressure in the bulb and capillary acting on the outside of the bellows. The greater pressure on the outside of the bellows is the result of saturation temperature plus superheat. The additional pressure on the outside of the bellows resulting from superheat, overcomes the spring loading which tends to close the valve.

A hand regulator is fitted for emergency use. It would be adjusted to give a compressor discharge pressure such that the equivalent condensing temperature shown on the gauge at the compressor outlet was about 7°C above the sea water

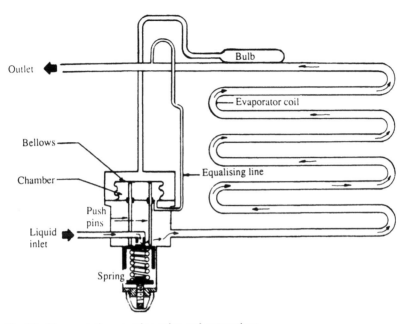

Fig. 50   Thermostatic expansion valve and connections

inlet temperature and the suction gauge showed an equivalent evaporating temperature about the same amount below the evaporator.

ROOM SOLENOIDS

The solenoid valve is opened when the sleeve (Fig. 51), moving upwards due to the magnetic coil, hits the tee piece and taps the valve open. It closes when the coil is de-energized and the sleeve drops and taps it shut. Loss of power therefore will cause the valve to shut and a thermostatic switch is used to operate it through simple on/off switching.

The thermostatic switch contains a bellows which expands and contracts under the influence of fluid in a capillary and sensing bulb attached to it. The

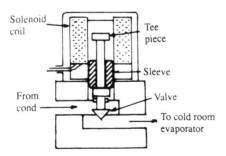

Fig. 51   Room solenoids

bulb is filled with freon or other fluid which expands and contracts with the temperature change of the space in which it is situated. As the temperature is brought down to the required level, contraction of the fluid deflates the bellows. The switch opens and the solenoid is de-energized and closes. Temperature rise operates the switch to energize the solenoid which opens to allow refrigerant through to the evaporator again. The switch is similar to the L.P. controller.

LOW PRESSURE CONTROLLER

The low pressure control stops the compressor at low suction pressure caused by closure of all cold compartment solenoids. When the pressure in the compressor suction rises due to solenoid opening, the L.P. control restarts the compressor.

The controller shown (Fig. 52) is of the Danfoss type operated through a bellows. The spring on the left is for the stopping pressure and that on the right for starting. The push pin operates the switch through a copper plate with a coiled spring between the two tongues. With the contacts open the spring is

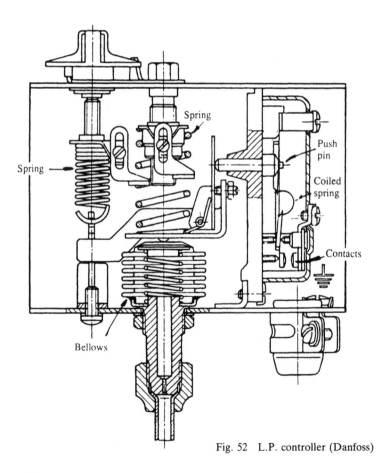

Fig. 52   L.P. controller (Danfoss)

coiled as shown. Outward movement of the pin compresses the spring and finally tips it to the position where the contacts close.

CONDENSER

The shell and tube sea water cooled condenser, for a freon system, is much the same as tubular coolers described in the heat exchange section. Materials are those suitable for resisting sea water corrosion. Heat transfer is sometimes improved by rolling threads on the outsides of the tubes.

EVAPORATOR

Each cold room used for domestic stores will have an evaporator which cools air blown through it by a fan. The air acts as a secondary refrigerant by circulating through and cooling the stores. Such an arrangement is termed a direct expansion system. Many cargo installations operate on the same principle.

There is a risk with direct expansion that a leakage of gas will occur into the cargo space and brine is used to circulate the air cooling coils in some cargo spaces. The evaporator is then a brine cooler and the brine is pumped through the air cooling grids in the cold rooms.

COMPRESSOR

Reciprocating compressors for systems cooling domestic store rooms are of the vertical in line type. For large cargo installations the banks of cylinders are arranged in V or W configuration as in the example shown.

Each crank carries the bottom ends of the four pistons in the cylinders. In older machines, pistons are of cast iron while modern compressors have aluminium alloy pistons. Piston rings may be of plain cast iron but special rings having phosphor-bronze inserts are sometimes fitted; these assist when running in. Connecting rods are H section steel forgings with white metal lined steel small end bushes. Liners are of high tensile cast iron and the crankcase and cylinders comprise of a one piece iron casting. The two throw crankshaft is of spheroidal graphite cast iron. Each throw carries four bottom ends as mentioned above but in other machines the number of banks of cylinders may be less. Main bearings are white metal lined steel shells.

Gas from the evaporator passes through a strainer housed in the suction connection of the machine. This is lined with felt to trap scale and other impurities scoured from the system by the refrigerant during the running in period. Freons are searching liquids, being similar to carbon tetrachloride. Thus they tend to clean the circuit but the impurities will cause problems unless removed by strainers. Any oil returning with the refrigerant drains to the crankcase through the flaps at the side of the cylinder space.

The valve assembly is shown in Fig. 54 in more detail. The delivery valve is held in place by a safety spring which is fitted to allow the complete valve to lift in the event of liquid carry over to the compressor. The delivery valve is an annular plate with its inside edge seated on the mushroom section and its outside edge on the suction valve housing. The suction valve passes gas from the suction space around the cylinder.

For unloading a mechanism holds the valve open so that gas is able to flow freely in and out through the valve without compression. The collar holding the

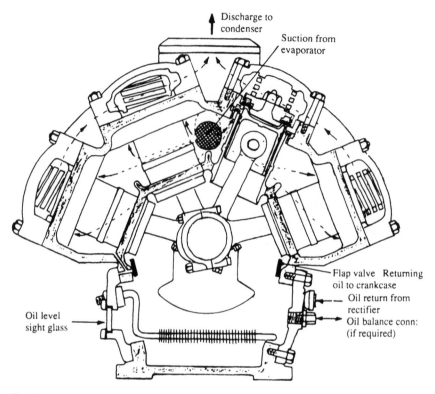

Fig. 53   Freon compressor

fingers is fitted around the liner and moved up or down by a yoke operated from a cam or servo cylinder.

The control system includes a high pressure cut-out but a safety bursting disc is also fitted between the compressor discharge and the suction. This may be of nickel with a thickness of 0.05 mm. A ruptured disc is indicated by suction and discharge pressures being about equal.

SHAFT SEAL

Where motor and compressor casings are separate, a mechanical seal is fitted around the crankshaft at the drive end of the crankcase. This prevents leakage of oil and refrigerant from the crankcase. The type shown (Fig. 55) consists of a rubbing ring with an oil hardened face against which the seal operates. The seal is pressed on to the face by the tensioning spring and being attached to a bellows, it is self adjusting. The rubbing ring incorporates a neoprene or duprene ring which seals it to the shaft.

The mechanical seal is lubricated from the compressor system and can give trouble if there is insufficient or contaminated oil in the machine. Undercharge

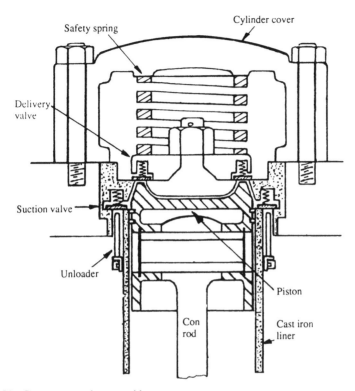

Fig. 54 Compressor valve assembly

may be caused by seal leakage due to oil loss. When testing the seal for leakage the shaft should be turned to different positions if the leak is not apparent at first.

LUBRICATION

Oil is supplied to the bearings and crankshaft seal by means of a gear pump driven from the crankshaft. The oil is filtered through an Auto-Klean strainer and/or an externally mounted filter with isolating valves. A pressure gauge and sight glass are fitted and protection against oil failure is provided by a differential oil pressure switch. Oil loss from the compressor is sometimes the result of it being carried into the system by the refrigerant.

Oil pressure is about 2 bar above crankcase pressure and the differential oil pressure switch is necessary to compare oil pressure with that of the gas in the crankcase. There is a relief valve in the oil system set to about 2.5 bar above crankcase pressure.

Cylinder walls are splash lubricated and some of the oil is carried around with the refrigerant. A float controlled oil trap (Fig. 56) may be fitted to reduce carry

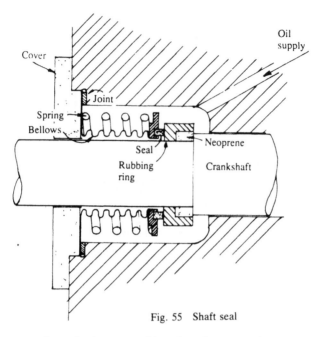

Fig. 55 Shaft seal

over from the larger machines but they are not always considered necessary. When used, the oil trap is fitted in the discharge pipe close to the compressor. The heat of the freon must be retained to prevent it from condensing and returning with the oil to the compressor crankcase. Thus the casing is insulated. Oil and refrigerant enter the trap by a pipe which runs down towards the bottom. The gas leaving this pipe changes direction but the heavier oil tends to drop to the bottom. The float rises as the oil collects, lifting a valve which allows

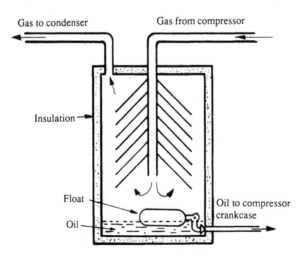

Fig. 56 Oil trap

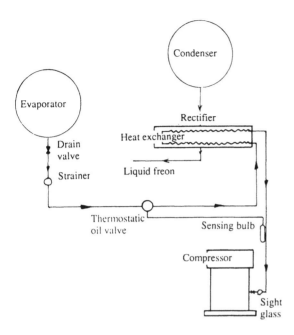

Fig. 57   Oil rectifier

the oil to be fed back to the compressor sump. Operation is improved by fitting a demister or scrubber unit.

In some installations there may be a problem caused by a tendency for oil to collect in the evaporator under certain conditions such as at low load when the speed of movement and agitation of the evaporating refrigerant are insufficient to keep the oil moving. To prevent loss of oil from the sump to the system, an oil rectifier (Fig. 57) may be fitted. The oil is automatically bled from the evaporator to a heat exchanger in which liquid refrigerant mixed with the oil is vapourized. The heat for vapourizing the refrigerant is obtained by passing warm liquid freon from the condenser, through the heat exchanger. Vapour and oil are passed to the compressor where the oil returns to the sump while the freon passes to the compressor suction. The regulator is a thermostatically controlled valve which operates in the same way as the expansion valve on the main system. It automatically bleeds the oil from the evaporator so that the gas leaves the rectifier heat exchanger in a superheated condition.

REFRIGERATOR OIL

The lubricant is required only in the compressor where its temperature will be in the region of 50°C. It must have the correct viscosity, film strength and stability both chemical and thermal for the operating conditions. Oxidation is not a problem because the system is freon filled. Water contamination may cause the oil to emulsify. With some refrigerants, water contamination produces acidity and corrosion.

Oil is also carried over to the low temperature side of the system where it must not interfere with heat exchange by congealing in the evaporator. It must not

73

form deposits on control valves and in passages, which would also interfere with the operation of the system.

Low pour point is required. Some oils have a lower pour point than others due to their nature. These are the naphthenic oils which have generally a lower pour point than the corresponding grades of paraffinic oils. Either may be used because they are refined and dewaxed to the lowest pour point possible. Pour point depressants are also used as additives in the oil.

Freon or refrigerant 12 is miscible with oil at the working temperatures and pressures in the system. The pour point of the mixture tends to be lower than the evaporator temperature. Oil pour point therefore is less important in an installation using refrigerant 12 than in those using a refrigerant which is not miscible such as CO_2. Because the oil is in solution it does not settle out on to surfaces. When the refrigerant boils off in the evaporator the agitation and velocity of the gas carries the oil mist along to the compressor suction. Refrigerant 22 is miscible with oil in the condenser but in the evaporator cold conditions there are two liquid layers the top mostly oil and the bottom mostly refrigerant.

Floc point is a term applied to oils in freon systems. Cooling of oil in solution with freon causes a wax to precipitate initially to produce a cloudy appearance and finally as crystals of wax. In this state the wax is called a flocculant, hence the term floc point. Tests for wax precipitation are carried out with a 10 per cent solution of oil in freon which is cooled until wax appears. The temperature at which this occurs, is the floc point. Flocculation in a system can cause wax to deposit on regulating valves and interfere with operation.

Pour point applies to all oils but floc point to oils in freon systems.

Oil viscosity is important because of the variation produced by miscibility with the gas.

In general, the oil used must be compatible with the refrigerant. It is a straight mineral with additives to prevent foaming and to inhibit against chemical action with refrigerants and system metals. A naphthenic oil may be used for low temperatures. Oils are de-waxed and refined for low pour point and floc point. Because the system must be moisture free, it is important that oils are supplied with no water content.

AIR IN THE SYSTEM

This is indicated by an abnormally high condenser pressure gauge reading and possibly by the presence of small bubbles in the sight glass.

Remedy The procedure for removing air from the system, is similar to that used for removing excess refrigerant. The condenser liquid outlet is closed and with the circulating water on, the charge is pumped to the condenser and receiver. The air will not condense but remain in the top of the condenser above the liquefied refrigerant. It causes the pressure to be higher than normal. The air is expelled by slackening the purge valve. After the air has been removed, the valves are reopened and the machine restarted. Refrigerant is added if necessary.

MOISTURE IN THE SYSTEM

Water circulating with the freon tends to freeze on the regulator causing a build up of pressure on the condenser side and drop in pressure on the evaporator side due to the blockage. The machine tends to be stopped by the high pressure cut-out.

Remedy Driers are used to remove moisture, and icing of the expansion valve would indicate that the chemical is no longer effectively removing the moisture. The chemical (either activated alumina or silica gel) is renewed and the compressor restarted after the ice has melted. Usually the ice in the expansion valve will melt due to the ambient temperature.

UNDERCHARGE

Symptoms of undercharge are a low condenser gauge reading and the appearance of large bubbles in the sight glass. The compressor will tend to work hot also. The result of undercharge is that the performance falls off.

When undercharge is suspected a leak test is necessary to confirm and locate the fault. For freon systems a leak detector lamp burning methylated spirits or bottled gas can be used. The lamp has a pale blue or colourless flame which turns to green when freon is drawn into it by a sampling tube. The open end of the tube is held close to joints and other potential leakage points around the pipework. The mechanical seal on the shaft is tested because freon from the crankcase may be lost through wear or a fault. A large leak will cause the flame to burn violet and it is sometimes necessary to ventilate the space to clear excess gas before the leak can be pinpointed.

If no lamp is available, escaping refrigerant can be detected by brushing soapy water over the joints and flanges. Bubbles indicate the leak (a soap test was used on the old CO_2 machines).

Remedy When adding gas, the freon storage cylinder is connected to the filling valve on the regulator outlet, loosely. The bottle valve is cracked open to clear air from the connecting pipe and the nut is tightened on the nipple of the filling valve. The charging connection is made to the liquid stop valve or suction stop valve if there is no valve after the regulator. The bottle must be kept upright to prevent entry of liquid when the connection is made to the suction side of the system. The charging valve is opened to one turn off the back seat and with the compressor running, the bottle valve is fully opened. Charging is continued until the bubbles disappear from the sight glass. Charging will correct the pressure gauge readings (i.e. condenser gauge about 7°C above the sea water inlet and suction gauge about 7°C below the evaporator on the equivalent saturation temperatures for the pressures).

OVERCHARGE

Overcharge is indicated by a high condenser gauge reading with a full liquid sight glass.

There are other faults which will result in high condenser pressures. These include poor cooling, having air in the system and icing of the regulator. Such faults occur spontaneously but overcharge only happens when the system has been charged.

Remedy The charge is pumped to the condenser and the excess refrigerant is released to atmosphere through a pipe connected up for this purpose.

When pumping the charge to the condenser, the cooling water is left on so that the gas from the compressor will be liquefied. The high pressure cut-out will stop the compressor, as the pressure rises. Condenser cooling becomes less effective with accumulation of liquid and at the end of the process it will be necessary to restart the compressor by hand in order to fill the condenser.

OTHER OPERATIONAL FAULTS

The cooling water can be restricted due to choked sea water pump strainers or due to chokage of the system; the supply may be reduced due to a pump fault. The effect of the poor cooling will be that the refrigerant will not be efficiently liquefied and the condenser pressure gauge will show a high reading due to the excess of gas. The condenser ends and pipes will feel hot.

Heat exchange would be reduced by oil deposits in condensers (and evaporators) but this is not usually a problem when suitable oils are used. Oil carry over is cut down by the separators fitted on some compressor discharges. Oil deposit is removed by chemical cleaning as described in the heat exchange section.

Frost on the evaporator coils reduces the efficiency of the plant by acting as an insulator between the evaporator and the air in a direct expansion system. The air flow is also restricted by the blockage. Automatic defrosting keeps the coils free of ice but failure of the defrost arrangement allows excessive icing. The result is that cold room temperature gradually rises and the compressor runs continuously at first. Later the compressor will cut out as the result of low suction pressure and then restart as the refrigerant passing through the still open solenoid, builds up pressure. Suction pressure is low because the thermostatic valve, controlled by the evaporator gas outlet temperature (due to icing this will be low) will reduce refrigerant flow.

Ice on the evaporator can be removed by washing it off with a hot water hose (with the plant shut down) after clearing the drip tray drain if necessary.

A blockage in the system may be caused by moisture forming ice on the expansion valve but it can be due to blocked strainers, closed valves or solenoids which have failed to open.

Electrical faults are responsible for a large number of refrigerator problems. Ship vibration, in turn, is the reason for many of the electrical faults. Loose connections and broken wires and earths, resulting from chafed insulation, are examples. The wiring diagram for the installation should be available, for location of all fuses etc.

Short cycling is the term used to describe a compressor unit repeatedly running for a few seconds and then cutting out. This is the result of operation of the L.P. controller. The control is arranged to operate when all solenoids have closed and suction pressure drops, to stop the machine. It restarts it when the suction pressure rises from solenoids reopening. Thus any condition which varies suction pressure over this range will cause the compressor to cut in and out.

If a solenoid is opened in the normal way by high cold room temperature, the refrigerant in passing through will build up suction pressure and the compressor will be started. If the supply is restricted and insufficient for compressor demand, the suction pressure will drop and the L.P. controller will stop the machine. The thermostatic valve may be the restricting device, due to low superheat of the gas leaving the evaporator (from the insulating effect of ice on the coils) it will be closed in to reduce the refrigerant flow.

AIR CONDITIONING

Before the installation of air conditioning as a normal practice, steam heating coils were fitted in the ventilation system to give some measure of heat, when required, to the air. Radiators were an additional or alternative heat source. With some exceptions, there was no attempt to cool air, but circulation by the ventilation fans was beneficial. Dry and moving air will take up perspiration as a vapour which removes latent heat from the skin as it evaporates. Ventilation also cleared away stale air and prevented inside temperature conditions from getting worse than those outside.

Straight heating of air from the outside when the ambient temperature is very low, to attain an inside comfortable level of about 21°C, greatly increases the capacity of that air to take up moisture from any available source. The hot air tends to dry the nasal passages, mouth and throat and cause discomfort, unless a means to humidify it is provided.

The SINGLE DUCT AIR CONDITIONING unit shown (Fig. 58) can provide both heating and cooling with control of humidity. When neither heating or cooling are required the plant is operated as a ventilating system only.

Heating for the air is provided by three steam heating coils, one for each space. Each is controlled by a thermostat set to give a temperature in the space of about 21°C. Individual section thermostats are necessary to maintain even temperature in different areas which are affected by other factors. Accommodation space near the engine room needs less heating than that in the upper part of the ship, for example. About 25 to 30% of the air is drawn from the outside, the balance is recirculated. This saves heating cost but still provides a freshening supply and makes up for losses.

When the outside temperature is cold, but not excessively so, the steam heater alone is used to maintain accommodation conditions. If outside temperature is very low then heat loss from the spaces will be high and both the recirculating and fresh air will tend to have a drying effect due to the considerable heating. Extra moisture added to the air by the humidifier reduces its drying effect. For comfortable conditions there should not be less than 40%

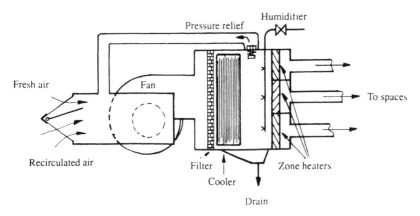

Fig. 58   Single duct simple air conditioning system

relative humidity when the accommodation is at 21°C. The valve controlling steam for the humidifier may be hand-operated. It must be closed when the air is not being heated or when the fan is stopped.

Limiting humidity to no more than the minimum 40% in very cold conditions will reduce condensation on inside surfaces of the external bulkheads.

Cooling in the simple unit shown is by a freon direct expansion plant. The mixture of fresh and recirculated air is delivered via the evaporator, where it is cooled, to the spaces served by the fan. Local temperature is adjusted by volume control at the delivery point.

The evaporator removes some of the heat that sustains moisture in suspension as vapour in the air. If the air carries a lot of moisture (has high relative humidity) cooling will bring it to the dew point so that moisture is precipitated, and it will continue to condense out as the temperature is further dropped. At the end of the air's passage through the cooler it will have lost moisture (been de-humidified) but be left with the maximum moisture it could carry at the new low temperature: its relative humidity would be 100%. Such air would be unable to absorb further moisture. Perspiration, instead of evaporating to cool the skin would remain as unpleasant wetness, while the heat would make people perspire more. The problem can be overcome by over-cooling the air in the cooler, so de-humidifying it to a greater extent (i.e. removing more water) and then reheating slightly to bring temperature to the comfort level. Reheating increases the capacity of the air for carrying moisture and therefore drops its relative humidity. Final temperature is 21°C (higher if outside air is very hot) and relative humidity about 50%, for comfort.

Condition of air in the spaces is checked with wet and dry bulb thermometers, actual temperature being obtained from the dry-bulb instrument. Water in wet muslin around the other thermometer bulb evaporates to a degree which is governed by the moisture content of the surrounding air. The evaporation takes latent heat from the bulb, causing it to be colder. The temperatures registered are used to find, from a psychrometric chart or table, the relative humidity. In general, whether the air conditioning system is used for heating or cooling a temperature of about 21°C and relative humidity of 50% is comfortable. Humidity is set lower in very cold conditions; temperature higher in very hot weather.

Nylon filters are provided to keep the air clean (removed for washing every six weeks) and the drain clears excess water from humidification or de-humidification. Extraction fans discharge air from spaces such as the galley and toilets to the outside. This reduces air pressure in these areas so that tainted air will not flow from them to other spaces but any flow will be in the other direction.

Other systems Heating of the air may be by steam (as above), hot water circulation or electric heating elements. Cooling by chilled water or brine may be used instead of direct expansion. A humidistat can be installed for automatic humidity measurement and control.

TWIN DUCT SYSTEM This gives the greater flexibility of temperature and ventilation required in a large passenger vessel (Fig. 59). One set of ducting carries warm air, the other set carries a cooler supply from a central air conditioning unit to the accommodation spaces. The cold/warm air ratio is controlled within a particular cabin or compartment by a local mixing unit.

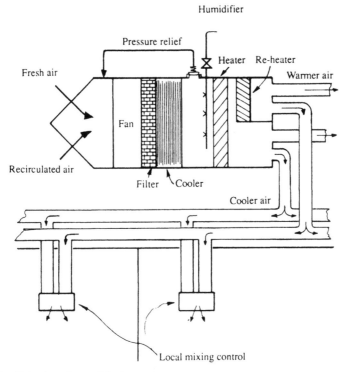

Fig 59 Twin duct air conditioning system

Two temperatures are produced in the air conditioning plant by using the reheater on a proportion of the air supplied.

LOCAL REHEAT SYSTEM Another variation uses an air conditioning unit with single duct type distribution and local reheating at the outlet in the space served. Individual temperature requirements are met by an electric element or hot water heat exchanger controlled by a locally set thermostat.

LEGIONELLA BACTERIA

Legionella bacteria is a type of pneumonia which may be fatal to older people, and its presence has been associated with the air conditioning plant of large buildings. The outbreak which led to investigation occurred at a convention for American ex-servicemen (the American Legion) and the identified cause of the problem was therefore labelled *legionella bacteria*.

There is a risk that the bacteria could flourish in the air conditioning systems of ships and consequently a Department of Transport M Notice has been issued to give warning and to recommend preventative measures.

The M Notice explains that the organisms breed in stagnant water or in wet deposits of slime/sludge. Possible locations for bacteria colonies are mentioned as being at the air inlet area and below the cooler (stagnant water), in the filter, in humidifiers of the water spray type and in exposed insulation. Provision of adequate drainage is recommended to remove stagnant water. Guidance is given for weekly inspection and cleaning as necessary of filters with a 50 p.p.m. super-chlorinated solution; and for the solution to be used on the cooler drain area at not more than three month intervals. Regular sterilization is called for with water spray type humidifiers (steam humidifiers being preferred).

Reference

Merchant Shipping Notice no. M1215 (1986). *Contamination of Ships' Air Conditioning Systems by Legionella Bacteria.*

CHAPTER 6

Metallurgical Tests

NON-DESTRUCTIVE TESTING (N.D.T.)

Non-destructive tests are carried out on components, not test pieces. They are used to detect flaws or imperfections during manufacture or those that develop during service. The tests give no indication of mechanical properties.

Visual inspection for surface defects is assisted by penetrant or magnetic crack detection to find the presence and full extent of hairline cracks. Where internal flaws are suspected, use is made of X-rays or ultrasonic testing. There are special devices for examination of machine finish.

LIQUID PENETRANT METHODS

One type of test uses a low viscosity liquid, containing a fluorescent dye. The area to be tested is sprayed or soaked and after a time lapse, to allow for penetration by capillary action, is wiped dry. When viewed under the ultra violet light, any faults will be shown up by the glow of the penetrant in them.

Another test uses a penetrant containing a powerful dye. This is sprayed on the suspect area with an aerosol. After allowing time for penetration, the area is wiped clean and covered with a liquid which dries to leave a chalky sediment (developer). The penetrant stains the developer along the line of the crack.

These methods are based on the old chalk and paraffin tests but the penetrants can have a hydrocarbon or alcohol base. Some are emulsifiable for removal by water spray, others can be cleaned off with solvents to reduce possible fire risk.

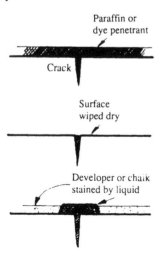

Fig. 60   Crack test with liquid penetrant

MAGNETIC CRACK DETECTION

This type of test is suitable only for materials which can be magnetized (can not be used for austenitic steels or non-ferrous metals). After the test the component is normally de-magnetized.

A magnetic field is produced in the component by means of an electric current or permanent magnet (Fig. 61) and magnetic particles are spread on the surface. Cracks are revealed by a line of magnetic particles.

The powder used, may be black iron oxide held in suspension in thin oil. It is poured on to the surface, the surplus being collected in a tray beneath. Coloured magnetic inks in aerosols are also available and the dry method makes use of powder only and this is dusted on the surface. Powder tends to collect at a crack in the same way as iron filings will stick to the junction of two bar magnets, placed end to end with opposite poles together.

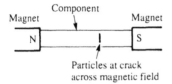

Fig. 61   Magnetic crack test

RADIOGRAPHIC INSPECTION

X-rays and gamma rays are used for inspection of welds, castings, forgings etc. Faults in the metal affect the intensity of rays passing through the material. Film exposed by the rays gives a shadow photograph when developed.

There is a requirement for radiographic examination of many welds, particularly those in pressure vessels. The sketch shows the methods for examining longitudinal and circumferential welds (Fig. 62).

Defects such as porosity, slag inclusions, lack of fusion, poor penetration, cracks and undercuttings are shown on the film. Limits are placed on the extent

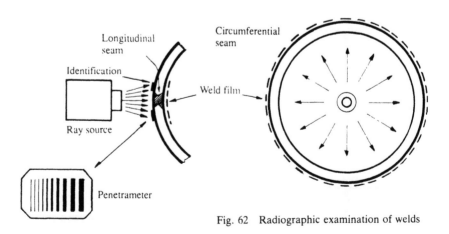

Fig. 62   Radiographic examination of welds

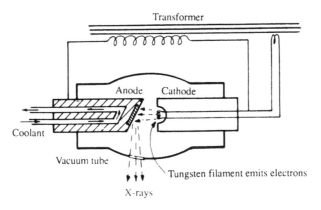

Fig. 63   X-ray machine

of defect, by the different inspecting bodies. The requirements may be given in booklets containing diagrams, charts and typical films.

An X-ray machine works on a similar principle to the thermionic valve. The electrons produced by the electrically heated filament in a vacuum tube are attracted by the positive anode (Fig. 63). The energy of the electrons striking the tungsten target as they are accelerated across by the high voltage, is mostly converted to heat, which is absorbed by the copper and removed by the coolant. About 1 per cent of the electrons are deflected as X-rays through the side of the tube.

A transformer is used to obtain the high voltage. In general, voltage ranges from 200 kV up to 400 kV. Increase of voltage produces rays of shorter wavelength which have greater penetration. Wavelengths are below those of light, radio or heat and in the order of 10^{-8} to 10^{-10}cm. They may be referred to in angstrom units ($1A = 10^{-8}$cm). Intensity of rays is measured in roentgen (r). This unit is defined as the quantity of ray energy which in passing through 1 cc of dry air at 0°C and pressure of 1 atmosphere, releases by ionization a quantity of electricity equal to one electrostatic unit.

The alternating current supplied by the transformer is rectified because the electrons will only move to the tungsten target when it is anodic to the filament. The current flow is measured in mA and gives an indication of ray intensity.

Gamma rays are produced by a radioactive source such as cobalt 60, iridium 192 or caesium 137. They are an alternative to X-rays. They have a shorter wavelength and are spontaneously emitted by decay of the source. Each source has an associated wavelength and a different rate of decay which gives a loss of intensity with age. Rate of decay is given as the time to halve the radioactivity and is known as the half life. Loss of intensity must be considered when calculating exposure time, which varies also with the different thicknesses and densities of materials.

The sealed capsule for the gamma ray source (Fig. 64), protects personnel from harmful radiation and the radioactive source from damage in transit. The gamma ray source is easily portable for use on site. Neither an electrical supply nor cooling are required.

Films from radiographic examination provide a permanent record of quality

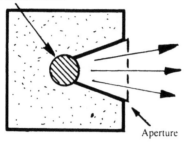

Radio-active material

Aperture

Protective lead box

Fig 64   Gamma ray source

of welds etc., and must be identified by serial numbers or other locating marks. Image quality indicators are placed on or adjacent to welds. There are various types and the wire type (Fig. 62) consists of a number of wires across the weld. Sensitivity is given by the percentage ratio of diameter of the thinnest wire visible, to maximum weld thickness. The lower the figure, the higher the sensitivity. Two per cent is required by the D.Tp. Obviously faults of the size of the thinnest wire that can be seen on the film will also be visible.

Radiographs are viewed by a radiologist on a uniformily illuminated diffusing screen. Training is necessary for the interpretation of film both with regard to the faults in the part being examined and misleading marks that sometimes appear on film.

A skilled radiographer is required for obtaining photographs. Exposure times for gamma rays vary with the type of material, its thickness and the intensity of the rays. X-ray machine voltage and exposure time are also varied to suit the material and its thickness. Distances between ray source, faults and film are important for image definition.

Rays are harmful either in a large dose or a series of small ones where the effect is cumulative. Monitoring against overdose is necessary with film badges, medical examination and blood counts. Direct exposure is avoided by the use of protective barriers but there is a danger that objects in the ray path will scatter radiation.

ULTRASONIC TESTING

Internal flaw detection by ultrasonic means is in principle similar to radar. The probe emits high frequency sound waves which are reflected back by any flaws in the object. Reflections are also received back from the opposite surface. The probe is connected to a cathode ray oscilloscope which shows the results in a simple way.

A single probe can be used, which combines both transmitting and receiving functions. Alternatively separate devices for transmitting and receiving the sound signals, are available.

The probe contains a slice of quartz which is cut in a particular plane from a quartz crystal. The quartz has a special property which is that it will pulse if an alternating current is applied to it. Opposite faces of the crystal are coated with a thin metallic film for connection to the electrical supply. The quartz will expand

and contract at the rate of the applied frequency. The amplitude is greatest at resonant frequency (i.e. natural frequency of the quartz). The action is reversible in that mechanical pulses received by the quartz will produce a small current.

The slice of crystal is protected by a thin steel plate which carries the pulses. When the probe is in use, a smear of oil or grease acts as the contact between the material under inspection and the probe. The vibrations would not be carried satisfactorily across an air gap.

The cathode ray tube contains a hot cathode and tubular anodes which accelerate a stream of electrons so that they strike the oscilloscope screen, the inside of which is coated with zinc sulphide that fluoresces wherever it is touched by them. If the electrons are deflected slowly across the screen, the effect is to produce a line because the fluorescence lingers. The electrons are deflected by means of the metal plates (X_1X_2 and Y_1Y_2) built into the tube. The X plates will produce horizontal deflection when one is made positive to the other. The Y plates will produce a vertical deflection also due to positive charging. The positive plate attracts the negative electrons.

The oscilloscope contains a device which regulates the positive potential on the X plates in such a way that the electrons sweep slowly from X_1 to X_2 and then rapidly back to X_1. The slow sweep is the **time base**.

The quartz crystal (pulse generator) is triggered to give a short pulse of vibrations simultaneously with the start of the electron movement from X_1 to X_2. The pulse is fed to the Y plates causing a peak at the start of the horizontal line. Then the pulse of vibrations is reflected back by the opposite surface it generates an electrical signal in the quartz which is amplified and rectified when fed as a d.c. signal to the Y plates. This produces a peak at the end of the horizontal line. The distance part of these peaks measures the thickness of the material, taking into account the difference in scale.

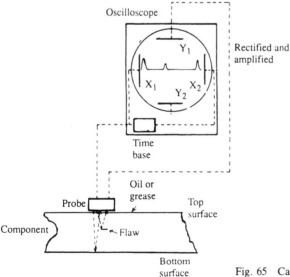

Fig. 65   Cathode ray oscilloscope

Any flaw in the material being inspected will also produce a peak. If the defect is large enough, it will show as a large peak at the expense of the peak at the right.

DESTRUCTIVE TESTING

Special test pieces are used which are damaged during the process.

TENSILE TESTING

The test pieces are machined to standard sizes depending on the thickness of the metal in question. When a material is tested under a tensile load, it changes shape by elongating. Initially the extension is in proportion to the increasing tensile load. If a graph is plotted showing extension for various loads, then a straight line is obtained at first (O to A in Fig. 66). If the loading is continued, the graph deviates as shown.

Within the limit of the straight line, if the load is removed the material will return to its original length. The graph can be plotted as load and extension or as stress and strain. Stress is load per unit area. Strain is extension divided by original length.

$$\text{STRESS} = \frac{\text{LOAD}}{\text{AREA}} \qquad \text{STRAIN} = \frac{\text{EXTENSION}}{\text{ORIGINAL LENGTH}}$$

Hooke's law states that within the elastic limit, stress is proportional to strain.

$$\text{STRESS} \propto \text{STRAIN}$$

Thus $\text{STRESS} = \text{STRAIN} + \text{CONSTANT}$

or $\dfrac{\text{STRESS}}{\text{STRAIN}} = \text{CONSTANT E}$

The constant is termed **Young's Modulus of Elasticity** and given the symbol E.

STIFFNESS

This is the property of resisting deformation within the elastic range and for a ductile material is measured by the Modulus of Elasticity (E). A high E value means that there is a small deformation for any given stress.

BEHAVIOUR OF THE MATERIAL

During the initial stretching of the test piece and until the elastic limit is reached, the cross sectional area reduces. It can be shown by experiment that a bar with the same elastic properties in all directions will have a constant relationship between axial strain and lateral strain. This is termed **Poisson's Ratio** and given the symbol ν.

$$\text{Thus POISSON'S RATIO} = \frac{\text{LATERAL STRAIN}}{\text{AXIAL STRAIN}}$$

At point Y the material yields i.e. there is a failure of the crystalline structure

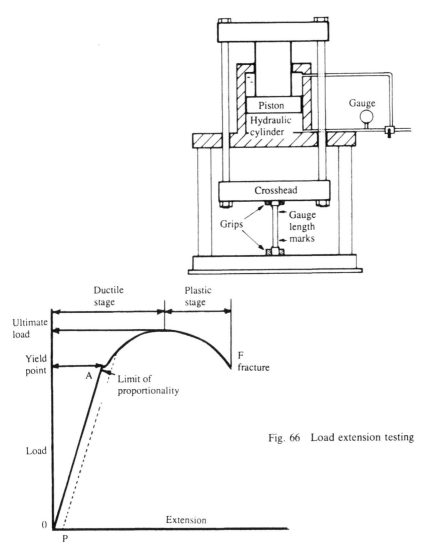

Fig. 66   Load extension testing

of the metal, not along the grain boundaries as has been happening, but through the grains themselves. This is known as slip. A partial recovery is made at the lower yield point, then extension starts to increase out of proportion to the load increase.

If the load is removed at any stage along the curve from Y to U, the material will be found to be permanently deformed by an amount OP. This is termed permanent set.

Maximum loading occurs at U and at this stage, local waisting or extension will start. Normally this starts at about the centre of the specimen and will rapidly be followed by failure.

ROUTINE TESTS

A full test is carried out for materials investigation. Acceptance tests for steel to be used in a pressure vessel or for weld test pieces etc., are based on ultimate tensile stress and percentage elongation. Sometimes reduction in area is required.

PROOF STRESS

For materials that do not have a marked yield point such as aluminium, there is a substitute stress specified. This is Proof Stress.

Proof stress is determined from a load/extension or stress/strain graph. (Fig. 67). It is obtained by drawing a line parallel to the straight portion and distant from it on the horizontal scale, by an amount representing a particular non-proportional elongation e.g. 0.1 per cent proof stress is found from a line through 0.1 per cent non-proportional elongation, as shown.

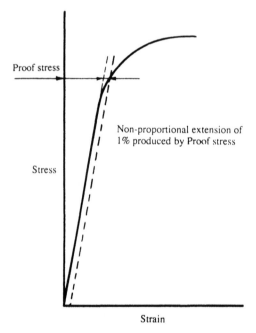

Fig. 67   Graph for proof stress

VALUES OBTAINED

If stress is plotted, then figures for Ultimate Tensile Stress, Yield Stress, Proof Stress and Breaking Stress can be read directly. If load is plotted then the loads at these points have to be divided by the cross-sectional area of the test piece to find the stress.

Percentage elongation is found from:

$$\frac{\text{EXTENSION} \times 100}{\text{ORIGINAL LENGTH}}$$

Percentage reduction in area is found from:

$$\frac{\text{REDUCTION IN AREA} \times 100}{\text{ORIGINAL CROSS-SECTIONAL AREA}}$$

FACTOR OF SAFETY

If a material is stressed beyond the elastic limit it will be permanently deformed. To prevent this, a factor of safety is used when calculating the working stress to which a material may be subjected.

$$\text{SAFE WORKING STRESS} = \frac{\text{ULTIMATE TENSILE STRESS}}{\text{FACTOR OF SAFETY}}$$

For steady loads, a safety factor of 4 may be used. Where there are shock loads, as in the drive chain for a camshaft, the factor of safety may be 25.

CREEP (HOT FAILURE)

Plain carbon steels when used at temperatures of 400°C and above tend to deform under stress. The rate of deformation is very slow and often is the result of loadings well below stress limits. In many instances the deformation takes the form of extension under tensile stress. This gradual change of shape due to steady stress is termed creep. It occurs in materials such as lead and tin, at room temperatures.

Steels used at high working temperatures in boilers, turbines and steam pipes etc., may have to be designed for lower stress than normal. Even in diesel engine cylinder heads, creep is thought to cause a gradual bulging upwards of the bottom of the cylinder head (the combined effects of heat and gas pressure being responsible).

Creep temperatures coincide with recrystallization temperatures of the various metals when changes in the crystal structure occur.

Alloys of steel have been developed which have creep resistance. Molybdenum is the essential alloying element and makes up about 0.5 per cent of the alloy steel but additions of small quantities of chrome and sometimes vanadium will improve creep strength.

SHORT DEFINITION

Creep is the slow plastic deformation of metals under constant stress and eventual failure at stress well below the normal failure stress.

Creep is an example of hot failure because fracture will ultimately follow the extension.

CREEP TESTING

Creep tests are carried out at controlled temperature over an extended period of time in the order of 10,000 hours. The test piece (Fig. 68) is similar to the type used for tensile tests and creep is usually thought of as being responsible for extensions of metals only. In fact creep can cause compression or other forms of deformation.

Temperature of the test is around that of recrystallization which for steels starts at about 400°C. For other metals the recrystallization temperature is different being about 200°C for copper and room temperature for tin and lead.

At the start of the test the initial load must be applied without shock. This load, normally well below the strength limit of the material, will extend the test piece slowly. The load is kept steady through the test and the temperature is maintained accurately.

Extension is plotted and the extension due to creep is seen to proceed in three stages. Initial and final extension periods are separated by a prolonged secondary stage of extension which follows a straight line law.

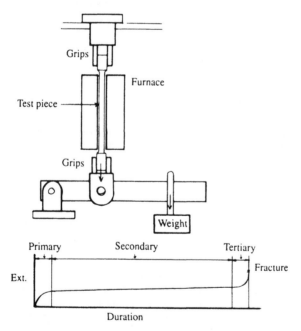

Fig. 68 Creep test

HARDNESS TESTING

The basis of the Brinell hardness test is the resistance of the material under test to deformation by a steel ball.

The equipment used for the test (Fig. 69) consists of a cylinder into which oil is forced by a pump. The oil pressure is raised until the top piston, which supports a weight, is floating. The bottom piston holds the hardened steel ball

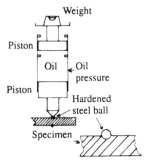

Fig. 69   Brinell hardness test

which is pressed into the metal beneath it. The loading for steel and metals of similar hardness is 3,000 kg. The load is allowed to act for 15 seconds to ensure that plastic flow occurs.

Surface diameter of the indentation is measured with the aid of a microscope which is traversed over the test piece on a graduated slide with a vernier. Cross wires in the microscope enable it to be accurately lined up above the edge of the indent. Knowing the ball diameter to be 10 mm, the surface area of the indentation can be calculated and the Brinell hardness number is found from the loading (3,000 kg) divided by the surface area of indentation. Hardness can usually be read from a chart once the indent diameter is known.

When the test is used on softer materials, the load is reduced. For copper it is 1,000 kg, and for aluminium 500 kg. Depth of penetration must be less than half of the diameter of the ball. Thickness of the specimen must be not less than 10 x depth of impression. The edge of the impression will tend to sink with the ball if the surface being tested has work hardened; otherwise the local deformation will tend to cause piling up of the metal around the indent.

If the above hardness test is used on very hard materials, the steel ball will flatten. This method is not reliable for readings over 600. It is used in preference to other methods where the material has large crystals, e.g. cast iron.

TYPICAL FIGURES

Mild steel has a Brinell number of about 130, cast iron about 200 and white cast iron about 400. A nitrided surface may have a hardness of 750. Non-ferrous metals vary.

IMPACT TESTING

Toughness of materials is compared by impact testing, not by tensile or other tests. Metals can have the same tensile strengths but different impact strengths.

A beam type test piece is used in the Charpy test (Fig. 70). The specimens for class 1 pressure vessel tests are of dimensions laid down and taken across the weld from the middle of the test plate. The test piece is laid across the supports with the notch on the opposite side from the impact point of the striker.

The striker is released and the swing of the pendulum after striking the test piece is used as an indication of impact strength. The tougher the material, the greater the amount of energy absorbed in fracturing it, and the smaller will be

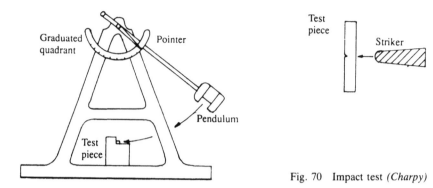

Fig. 70   Impact test *(Charpy)*

the extent of the swing after it has been fractured. The scale is kg.m, and the extent of the swing is shown by the small pointer.

Temperature is carefully controlled during impact tests and for materials which are to operate at low metal temperatures, the tests would be carried out at the temperature in question.

MACRO-SCOPIC EXAMINATION

The macro-specimen is taken from across the weld or any section that needs examination. It is polished and acid etched so that penetration and fusion are highlighted. Other defects may also become evident. A magnifying glass is used for closer scrutiny.

MICRO-SCOPIC EXAMINATION

It is possible to examine the crystals of a metal through a microscope if the surface has been finely polished and etched. The results of heat treatment and of metal failures are shown up in this way.

TESTS FOR CLASS 1 PRESSURE VESSELS

Boilers and air receivers are termed 'pressure vessels' and are manufactured by approved firms to the requirements of the various regulating bodies. The materials themselves must be from approved manufacturers and tested.

Weld tests   Test pieces are cut from the same material used for the shell plating. These are attached to the shell plate (Fig. 71) so that the longitudinal welds run through the test plates. The test pieces are treated in the same way as the pressure vessel material. However they may be straightened before being subjected to the same heat treatment as the pressure vessel. Usually test plates are not required for the circumferential seams.

A test plate is cut into different specimens with a piece in reserve for retests. Test piece (A) is of all weld metal. Gauge length, diameter and radius being in fixed proportions to each other. Diameter is limited by the thickness of the

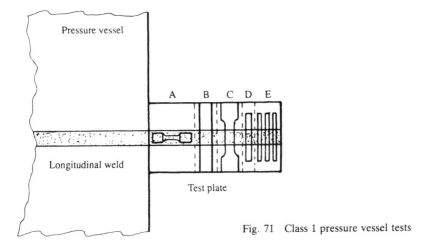

Fig. 71   Class 1 pressure vessel tests

metal. Results for tensile strength and elongation are required. Bend test specimens (B) are required for the inner and outer surfaces of the weld. The weld joint is tested by specimen (C) which is a tensile test piece taken across the weld. A macro specimen is taken at (D) and notch impact specimens from (E).

Radiographic examination of welds is required with lead type used for identification and image quality indicators. Ultrasonic examination is sometimes accepted as an alternative.

A hydraulic test is required on pressure vessels. A number of failures occur during pressure tests due to brittle fracture..These failures are frequently in pressure vessels of thick walled construction and occur in low temperature conditions.

FRETTING

A small relative movement between two metallic parts in close contact, can cause a form of mechanical wear, termed fretting. It is sometimes found on the backs of shell bearings and is due to slight movement of the bearing shell in its housing. Propeller shafts with a shrunk on liner and taper fitted propeller are sometimes prone to fretting troubles. The action occurs under the outboard end of the liner and under the forward end of the propeller hub. Fretting can occur in any area where there is a chafing action.

The wearing action produces, in turn, wear particles. These act as abrasives. In some cases the wear particles are found to be hard materials such as the chrome compounds from stainless steels.

Fretting corrosion   Metals are normally protected by an oxide film from corrosion. A fretting action exposes bare metal which then tends to oxidize. The wear particles themselves, tend to further oxidize. The appearance of the oxide like the red Fe_2O_3 from steels, is sometimes an indication that fretting corrosion is in progress.

Cracks occur in areas where there is fretting and extend due to fatigue. In the initial stages they can be removed by grinding. Areas of trouble are improved by relieving in the case of the fretting around shaft liners.

Brinelling is a form of fretting in ball races caused by vibration of an otherwise stationary race. A gradual indentation results in the race. The name is taken from the Brinell hardness test.

CHAPTER 7

Stern Tubes, Seals and Shafting Systems

The conventional midships position for the engines of older vessels, with the exception of tankers, was based on low engine power and strong hull construction. Shafts were long, but being of moderate diameter were able to flex with the hull as loading or other conditions changed (and in heavy weather). A loading or ballast condition which changed hull shape and shaft alignment to an unusual degree, sometimes caused higher temperature of some bearings due to uneven load distribution. Shaft stress was the hidden factor.

The trend towards higher engine powers and the positioning of engines aft, gave rise to large diameter, short length shafts of increased stiffness. Excessive vibration and resulting damage in many dry cargo and container vessels caused engines to be moved back towards midships (i.e., leaving one cargo compartment aft of the machinery space).

MATERIALS AND COUPLINGS

The intermediate shafting and the propeller shaft for a fixed propeller (Fig. 72) are of solid forged ingot steel and usually with solid forged couplings. Shafts are machined all over but of larger diameter and smooth turned in way of the bearings. The faces of flanged couplings are also smooth turned, with bolt holes carefully bored and reamered to give an accurate finish. Torque is transmitted by the friction between flanges and also through the shanks of the bolts. Each tightened bolt holds the flanges hard together in the area local to it. A circle of bolts is needed for a good all round grip. The design of flange couplings can be checked by formula given in Lloyds or other classification society regulations.

COUPLING BOLTS
(PROPELLER AND SHAFT FLANGES)

Elongation of a tensile test piece produces a related reduction in cross-sectional area. The behaviour of the metal is summarised by Poisson's Ratio, described in the previous chapter. The elongation of a bolt when tightened similarly causes a loss in area and diameter. In a clearance bolt this is not a problem, but with a fitted bolt positive contact between the accurately machined bolt and reamered hole would be affected. Shaft coupling bolts are tightened to force the faces of the flanges together so the friction between the faces will provide some proportion of the drive. However, fitted bolt shanks are also designed to take some load. A clearance bolt could provide the first requirement, but not the second. A fitted bolt when tightened and subject to reduction in cross-section would also fail on the second count and probably be damaged by fretting. A

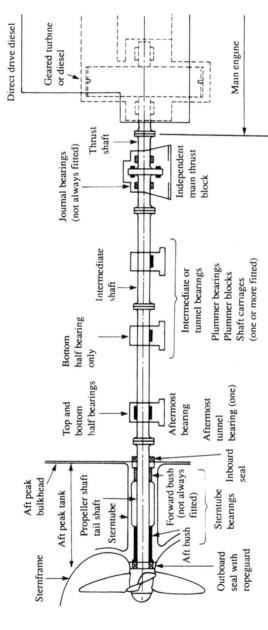

Fig. 72   Shaft system (Glacier)

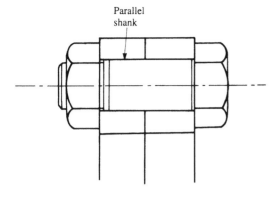

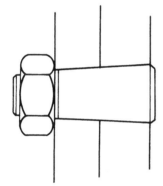

Fig. 73   Fitted bolt

Fig. 74   Tapered bolt

tapered bolt (Fig. 74) may be used instead of a conventional coupling bolt (Fig. 73) to obtain a good fit and the required tightening.

The Pilgrim hydraulic bolt uses the principle embodied in Poisson's Ratio to provide a calculated and definite fitting force between bolt and hole. The bolt (Fig. 75) is hollow and before being fitted is stretched with hydraulic pressure applied to an inserted rod from a pressure cylinder screwed to the head of the bolt. Stretching makes the bolt diameter small enough for insertion into the hole after which the nut is nipped up. Release of hydraulic pressure allows the bolt to shorten so that (1) predetermined bolt load is produced and (2) diametrical re-expansion gives a good fit of the shank in the hole. These bolts, used in flange couplings and flange mounted propellers, have the advantage that they are easily removed for inspection and maintenance; also the problem of driving in is avoided.

MUFF COUPLING

An alternative to the conventional flange couplings for the tailshaft, the muff coupling allows the shaft to be withdrawn outboard. The SKF coupling (Fig. 76) consists basically of two steel sleeves. The thin inner sleeve has a bore slightly larger than the shaft diameter and its outer surface is tapered to match the taper

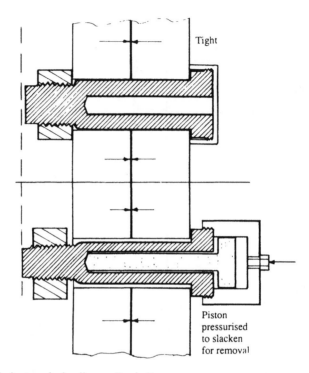

Fig. 75   Pilgrim type hydraulic coupling bolt

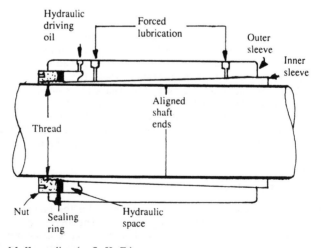

Fig. 76   Muff coupling (as S. K. F.)

on the bore of the outer sleeve. The nut and sealing ring close the annular space at the end of the sleeves.

When the coupling is in position, the outer sleeve is hydraulically driven on to the tapered inner sleeve. At the same time, oil is injected between the contact surfaces to separate them and thus overcome the friction between them. Oil for the operation is supplied by hand pumps; two for the forced lubrication and another hand or power pump for the driving oil pressure.

When the outer sleeve has been driven on to a predetermined position, the forced lubrication pressure is released and drained. Oil pressure is maintained in the hydraulic space until the oil between the sleeves drains and normal friction is restored. After disconnecting hoses, plugs are fitted and rust preventive applied to protect exposed seatings. A sealing strip is pressed into the groove between the end of the sleeve and the nut.

The grip of the coupling is checked by measuring the diameter of the outer sleeve before and after tightening. The diameter increase should agree with the figure stamped on the sleeve.

To disconnect the coupling, oil pressure is brought to a set pressure in the hydraulic space. Then with the shafts supported, oil is forced between the sleeves. The outer sleeve slides off the inner with a rate controlled by release of the hydraulic oil pressure.

SHAFT BEARINGS

The intermediate shafting (Fig. 72) if supported in plain or tilting pad bearings, has an aftermost bearing which is lined top and bottom. Roller bearings are installed in many vessels.

PLAIN AND TILTING PAD BEARINGS

The shaft supported in a plain journal bearing, will as it rotates, carry oil to its underside and develop a film of pressure. The pressure build up is related to speed of rotation. Thus oil delivered as the shaft turns at normal speed, will separate shaft and bearing, so preventing metal to metal contact. Pressure generated in the oil film, is effective over about one third of the bearing area (Fig. 77) because of oil loss at the bearing ends and peripherally. Load is supported and transmitted to the journal, by the area where the film is generated. The remaining two-thirds area does not carry load.

Replacement of the ineffective side portions of the journal by pads capable of carrying load, will considerably increase its capacity. Tilting pads based on those developed by Michell for thrust blocks (Fig. 78) are used for the purpose. Each pad tilts as oil is delivered to it, so that a wedge of oil is formed. The three pressure wedges give a larger total support area than that obtained with a plain bearing.

The tilt of the pads automatically adjusts to suit load, speed and oil viscosity. The wedge of oil gives a greater separation between shaft and bearing than does the oil film in a plain journal. The enhanced load capacity of a tilting pad design permits the use of shorter length or less bearings.

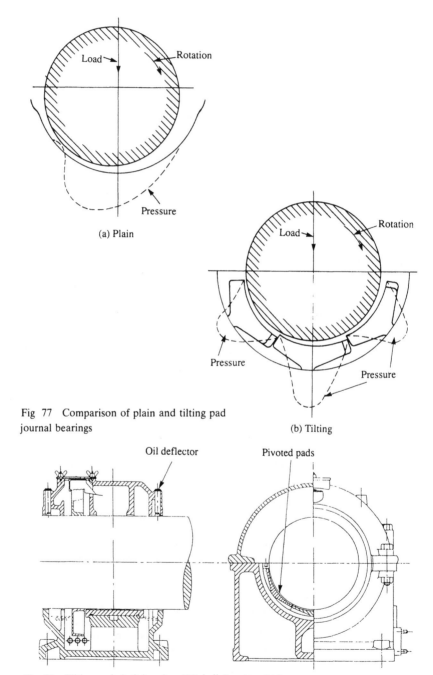

(a) Plain

Fig 77   Comparison of plain and tilting pad
journal bearings

(b) Tilting

Fig. 78   Tilting pad shaft bearing *(Michell Bearings Ltd)*

ROLLER BEARINGS

Roller bearings (Fig. 79) are supplied in sizes to suit shafts up to the largest diameter. Flange couplings dictate that roller bearing races must be in two parts for fitting.

The length of shaft where the split roller bearing is to be fitted must be machined very accurately and with good finish. The two halves of the inner and outer races are fitted and held with clamping rings.

Adequate speed for build up of fluid film pressure is vital for journal bearings.

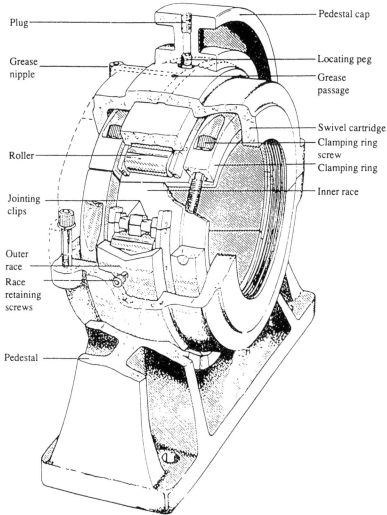

Fig. 79   Shaft roller bearing *(Cooper Roller Bearings Co. Ltd)*

At low speeds there may be metal-to-metal contact with wear and damage. Friction at low rotational speeds is high. Roller bearings are not dependent on speed for effective lubrication. Friction is low at all speeds. This makes them suitable for steam turbine installations and in ships where slow steaming may be necessary.

THRUST BLOCKS

The main thrust block transfers forward or astern propeller thrust to the hull and limits axial movement of the shaft. Some axial clearance is essential to allow formation of an oil film in the wedge shape between the collar and the thrust pads (Fig. 80). This clearance is also needed to allow for expansion as parts warm up to operating temperature. The actual clearance required depends on dimensions of pads, speed, thrust load and the type of oil employed. High bearing temperature, power loss and failure can result if axial clearance is too small.

A larger than necessary clearance will not cause harm to the thrust bearing pads, but axial movement of the shaft must be limited for the protection of main machinery.

The accepted method of checking thrust clearance involves jacking the shaft axially to the end of its travel in one direction and then back to the limit of travel in the other. Total movement of the thrust shaft (about 1 mm being typical) is registered on a dial gauge. Feelers can be used as an alternative between thrust ring and casing. Use of feelers in the thrust pad/collar gap is likely to cause damage and may give a false reading.

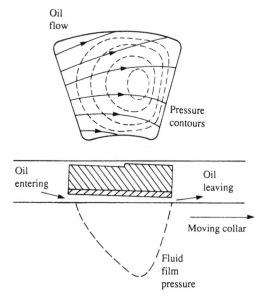

Fig. 80   Conventional Michell thrust pad

THRUST BLOCK POSITION

The siting of the main thrust block close to the propulsion machinery, reduces any problems due to differential expansion of the shaft and the hull. The low hull temperature of midship engined refrigerated cargo ships, caused a contraction relative to the shaft of, perhaps, 20 mm (¾ in). Variations can be caused by changes in water temperature or heating of fuel tanks. Other problems associated with the stern tube end of the shafting system include whirl of the tailshaft, relative movement of the hull and misalignment due to droop from propeller weight. Some thrusts are housed in the after end of large slow speed diesels or against gear boxes. Deformation produced by the thrust load can cause misalignment problems, unless suitable stiffening is employed (particularly with an end of gearbox installation).

THRUST BLOCK SUPPORT

The substantial seating provided for the main propulsion machinery provides an ideal foundation for the thrust block and a further reason for siting it close to the engine. The upright thrust block and any supporting stool must have adequate strength to withstand the effect of loading which tends to cause a forward tilt. This results in lift of the aft journal of the block (unless not fitted) and misalignment of the shaft.

Axial vibration of the shaft system, caused by slackening of propeller blade load as it turns in the stern frame or by splay of diesel engine crankwebs, is normally damped by the thrust block. Serious vibration problems have sometimes caused thrust block rock, panting of the tank top and structural damage.

THRUST PADS

The pivot position of thrust pads may be central or offset. Offset pads are interchangeable in thrust blocks for direct reversing engines, where direction of load and rotation changes. Offset pads for non-reversing engine and controllable pitch propeller installations are not interchangeable. Two sets are required. Pads with a raised central pivot position are interchangeable.

Some modern thrust blocks are fitted with circular pads (Fig. 81), instead of those with the familiar kidney shape. A comparison of the pressure contours on conventional kidney shaped pads and the circular type shows why the latter are effective.

STERN TUBES

The stern tube bearing is also part of the shaft support system. In some of the later designs, the bearing is accessible from the machinery space. The stern tube bearing being at the end of the shaft is affected by the overhanging weight of the propeller. The load pulls the outer end of the shaft down so that there is a tendency for edge loading of the stern tube bearing. The forward part of the tail or propeller shaft is tilted upwards. Weardown tends to make the alignment worse and whirl may give an additional problem.

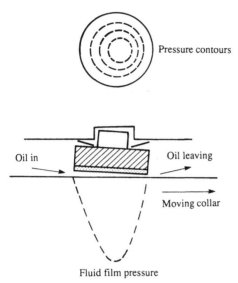

Pressure contours

Oil in

Oil leaving

Moving collar

Fluid film pressure

Fig. 81   Round thrust pad

SEA WATER LUBRICATED STERN TUBES

Sea water lubricated stern tubes (Fig. 82) are supported at the after end by the stern frame boss and at the forward end in the aft peak bulkhead. Their cast iron construction requires strong support in way of the bearing itself, from the stern frame boss. A steel nut at the outboard end retains the tube in position, with its collar hard against the stern frame and the bearing section firm within the stern frame boss. Welded studs hold the forward flange against the aft peak bulkhead. A brass ring secured with set screws and sealed with white lead protects the outer screw thread from sea water.

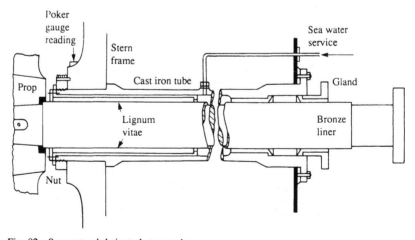

Fig. 82   Sea water lubricated stern tube

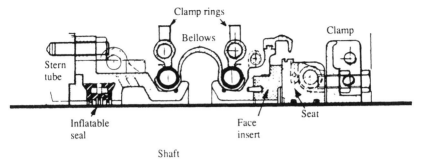

Fig. 83   Sea water lubricated stern shaft seal *(Maneseal)*

The traditional lignum vitae staves are fitted with end grain vertical beneath the shaft for better wear resistance. Staves in the upper part are cut with grain in the axial direction for economy. The staves are shaped with V or U grooves between them at the surface, to allow access for water. The grooves also accommodate any debris. They are held in place, in the bronze bush by bronze keys, attached to the bush by countersunk screws. Bearing length is equal to four times shaft diameter.

The alternative bearing materials are based on phenolic resins.

Sea water which enters at the after end or from the circulation system to cool and lubricate, is an electrolyte which will support galvanic corrosion. Wastage of the vulnerable steel shaft is prevented by a shrunk on bronze liner and rubber seal sandwiched between the propeller hub and the liner end. Sea water ingress to the machinery or tunnel space is minimized by the stuffing box gland.

Excessive weardown of bearing materials due to vibration or whirl, poor quality of work when rewooding, inferior materials, presence of sand/sediment in the water or propeller damage, could necessitate early rewooding. Bearing life for vessels with engines aft and particularly tankers and ore carriers which spend long periods in ballast has been short with rewooding being needed in perhaps eighteen months.

Radial face seals similar to those used for oil lubricated stern tubes can be used for sea water lubricated stern tubes (Fig. 83). These are fitted at the inboard end of the tube with the outboard end open. However, sea water is supplied from the sea water circulating system and runs out through the after end of the tube. The amount of any sand in the water would tend to be less after passing through the pipe system.

INSPECTION

During drydock inspection, bearing weardown is measured by poker gauge. Examination, after removal of the propeller and inward withdrawal of the propeller shaft (tailshaft), may reveal various defects (Fig. 84).

A keyway milled in the shaft taper acts as a weakening factor which allows some deformation of the surface. Transmission of torque from the shaft via the key, to the propeller hub causes a deformation which tends to open the keyway. Grip of the propeller along the side of the keyway does this as well. Cracks

105

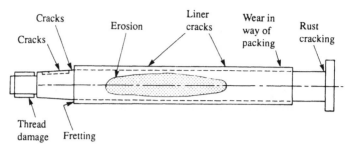

Fig. 84   Sea water lubricated propeller shaft defects

having been a problem, are reduced by the employment of sled type keys, radiused corners and spooning.

The rubber seal sandwiched by the propeller hub and protective bronze liner, prevents ingress of sea water which would act as an electrolyte to promote galvanic corrosion of the exposed part of the shaft. Wastage from corrosion or fretting of the steel shaft beneath the forward end of the hub or locally under the liner could weaken the shaft surface at the hub–liner notch to cause shaft failure through fatigue or corrosion fatigue. Shaft droop from overhanging weight of the propeller stretches the upper surface and compresses the lower, giving conditions when the shaft is rotating, which are likely to cause fatigue. The imposed alternating effect is of low frequency and high stress.

The shrunk on bronze liner, fitted to protect the steel shaft against 'black corrosion' may itself be damaged by working conditions. Shaft whirl can lead to patches marked by cavitation erosion, scoring occurs in way of the stern gland packing and liner cracking has sometimes penetrated through to cause corrosion cracking in the shaft.

OIL LUBRICATED STERN TUBES

Progress from sea water to early oil lubricated stern tubes involved exchange of the wood lined bronze carrier for a white metal lined, cast iron bush. Oil retention and exclusion of sea water, necessitated the fitting of an external face type seal. The stuffing box was retained in many early oil lubricated stern tubes, at the inboard end.

The later designs (Fig. 85) with an extended length boss built into the stern frame, provide better support for the white metal lined bearing. A minimum bearing length of two times the shaft diameter will ensure that bearing load does not exceed 0.8 N/mm² (116 lbf/in²).

The tube is fabricated and welded direct to the extension of the stern frame boss at the after end and to the aft peak bulkhead at the forward end.

Oil is contained within the Simplex type stern tube (Fig. 85) by lip seals. The elastic lip of each nitrile rubber seal, grips a rubbing surface provided by short chrome steel liners at outboard and inboard ends of the steel propeller shaft. The outboard liner additionally protects the steel shaft from sea water contact and corrosion.

Heat produced by the friction will result in hardening and loss of elasticity of the rubber, should temperature of the seal material exceed 110°C. Cooling at

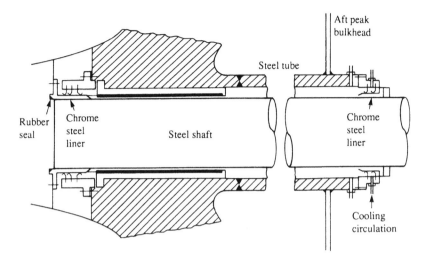

Fig. 85   Oil lubricated stern tube

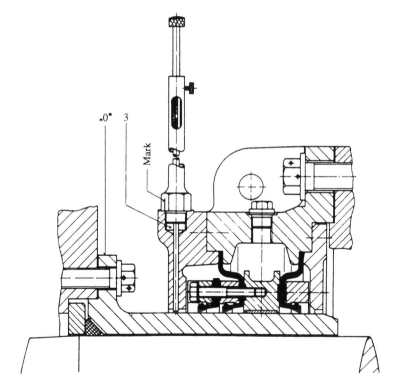

Fig. 86   Floating ring type lip seal assembly *(Simplex)*

the outboard end is provided by the sea. Oil circulation aided by convection, is arranged to maintain low temperature of seals at the inboard end. Connections are fitted top and bottom between the two inboard seals.

The chrome steel liners act as rubbing surfaces for the rubber lip seals and grooving from frictional wear has occurred. The problem has been overcome by using a ceramic filler for the groove or alternatively a distance piece to axially displace the seal and ring assembly. Allowance must be made for relative movement of shaft and stern tube due to differential expansion. New seals are fitted by cutting and vulcanizing in position.

Lip seals will accept misalignment but a floating ring design (Fig. 86) was introduced by one maker.

GLACIER-HERBERT STERN BEARING

The propeller shaft is flanged at the after end and the hub of the propeller is bolted to the flange. On the inboard side of the flange there is a carrier ring bolted. This forms a shroud around the spigot projecting aft from the spherical seating ring. Two inflatable seals with individual air supplies are fitted in the periphery of the spigot. These can be inflated to provide a seal against the inside of the carrier ring.

Sealing the stern bearing space, permits work to be carried out on the stern bearing and seals, without the necessity of drydocking. An alternative to using the inflatable seals, is to apply a bandage around the small gap between the carrier ring and the spherical seating ring.

The propeller shaft has two short rotating liners of chrome steel. The liner at the after end is bolted to the propeller shaft flange. The inboard liner is fixed by a clamping ring. These liners act as rubbing surfaces for the rubber seals. (The seals shown, are of the Simplex type, see Fig. 88). The outboard and inboard seal housings are attached to the spherical seating ring and the diaphragm, respectively.

The spherical carrier ring is bolted to the flange on the after end of the bearing tube, and is supported in the spherical seating ring.

The diaphragm is bolted to the flange on the inboard end of the bearing tube and is itself bolted to the stern frame casting.

'O' rings in the peripheries of the spherical carrier ring and of the diaphragm at the inboard end, seal the oil space around the bearing tube. The bearing tube is of spheroidal graphite cast iron and white metal lined. It is split along the horizontal and the two halves are bolted together through flanges along the horizontal join. Alignment of the bearing tube can be adjusted by the distance pieces and wedge chocks which hold the diaphragm. The axial bolts have constant loading due to the Belleville washer packs.

The stern frame is fully machined before being welded into the hull of the vessel. The bore for the spherical seating ring can be further machined for adjustment if necessary, but controlled welding is used to maintain alignment during hull construction. The shaft is installed from the outboard end, with its rotating liner and carrier ring assembled. The shaft inboard end is fitted with an oil injection coupling. This is the SKF shaft coupling, sometimes termed 'muff' coupling, described and illustrated above in Fig. 76.

The Glacier-Herbert stern bearing can be dismantled without drydocking, if necessary for maintenance or inspection.

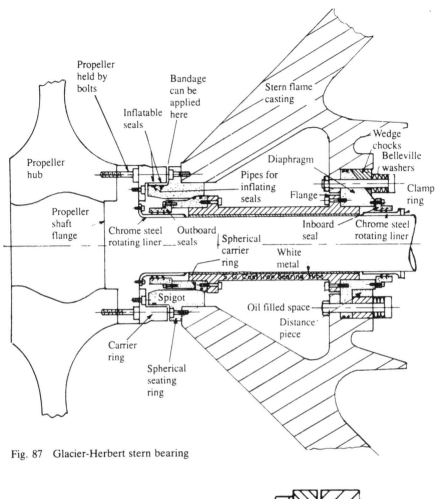

Fig. 87   Glacier-Herbert stern bearing

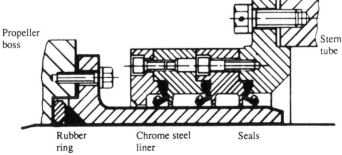

Fig. 88   Lip type of seal *(Simplex Compact)*

TURNBULL SPLIT STERN BEARING

The complete bearing consists of the section shown (Fig. 89) including the steel casting which is welded in to become part of the stern frame. A welding sequence is used to maintain bearing alignment during the operation. Installation time is reduced by this method as compared with that of boring out the stern frame. The shaft rests on a white metal lined cast iron shell in the housing. This forms the bottom half of the bearing. The top half bearing is secured by hydraulically operated jacks which are mechanically locked by nuts. The double acting jacks are also used to lift the cap clear of the shaft thus automatically bringing trolleys into contact with a built in overhead rail which enables the cap to be drawn forward.

Examination of the forward seal, bearing and tailshaft can be carried out afloat, at loaded draught, if necessary (Fig. 90). The lubricating oil is drained and the forward seal released and drawn forward. The top half bearing is lifted hydraulically and moved forward on the overhead rail. The tail shaft can be visually examined and crack detected back to the propeller flange. The full tailshaft surface is exposed by operating the turning gear. The alignment of the tailshaft relative to the bearing is checked by means of feeler gauges.

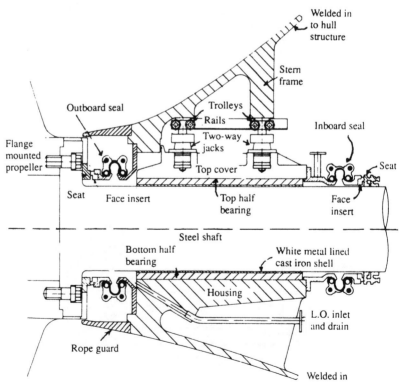

Fig. 89   Ross-Turnbull split stern bearing *(Mark I)*

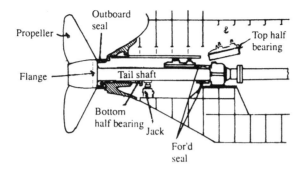

Fig. 90   Examination of Ross-Turnbull Mark I split stern bearing, showing removal of top half

Full inspection or replacement of the outboard seal and removal of the propeller mounting bolts for inspection and crack detection is carried out from outside the vessel after trimming or drydocking and removing the rope guard. Both seals are fully split being of Crane manufacture (see Fig. 93).

To examine or replace the bottom half bearing whilst the ship is afloat, the shaft is supported by a jack positioned forward of the bearing (see Fig. 90) and the bottom half bearing moved forward, rotated around the shaft and lifted clear.

An oil circulating system with a cooler is necessary for split stern bearings. Water in the aft peak has a cooling action on stern tubes which are oil lubricated but the accessibility of the split bearings means that they are situated in a void space. Oil pressure is kept higher than the external sea water pressure by a header tank arrangement or by a pneumatically pressurized compensating tank. The oil is pump circulated or may be caused to flow around the system by the pumping action of the bearing.

Instruments are fitted for readings of temperature, pressure, level and flow with alarms as necessary.

MARK IV BEARING   (Fig. 91)

Later versions of the stern bearing have a number of changes in design. These allow all maintenance and surveys to be carried out with the vessel in the water. Bearing alignment can also be adjusted.

The top half bearing is removed in much the same way as before (Fig. 92), after the inboard seal has been pulled forward. However, a shroud has been extended back from the stern frame casting so that it will act as a propeller support cradle. The casting also has a hydraulic/mechanical sealing ring, used to close the gap between the shroud on the forward end of the propeller boss and the stern frame. The position of the seal is such that the outboard working seal and the propeller flange bolts can be inspected or changed.

To carry out work on the outboard fittings, the maintenance seal (Fig. 91) is engaged hydraulically and mechanically locked in position. The space inboard of this auxiliary seal is drained before removing the bottom half bearing. The bottom half bearing complete with outboard working seal can then be with-

111

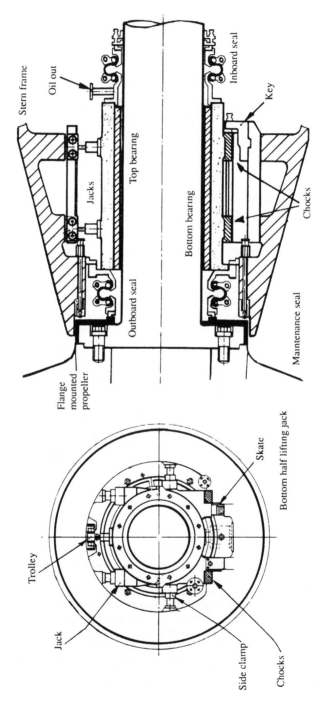

Fig. 91   Ross-Turnbull split stern bearing (*Mark IV*)

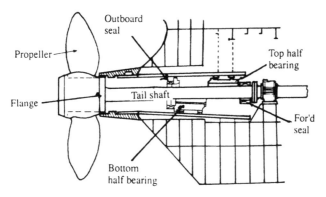

Fig. 92   Examination of Ross-Turnbull split stern bearing *(Mark IV Type)*

drawn into the vessel thus giving access to the shaft. After removal of the split seal which is bolted to the forward face of the tailshaft flange the propeller mounting bolts are exposed. These can be removed singly for inspection and crack detection.

When in working position, the bottom half of the stern bearing is clamped by jacks on either side. It is located by axial restraining keys and rests on chocks which are fitted after alignment. The top half bearing acts as the top clamp.

To remove the bottom half bearing (after clearing the top) the following procedure is used. Portable hydraulic jacks are placed under the bearing and the complete assembly of bearing, propeller and tailshaft is lifted so that the main chocks can be removed. Skates are placed under the bearing at the same time. With the chocks out, the assembly is lowered until the propeller is resting in the shroud which is part of the stern boss. Further lowering of the jacks brings the bottom half bearing away from the tailshaft and on to the machine skates. The jacks are removed and the bottom half bearing is brought forward on the skates, together with the seal face and bellows section of the oil seal.

The complete shaft is exposed for examination and crack detection. Removal of the split seal gives access to the bolts and dowels in the propeller flange, without disturbing the propeller.

CRANE SEALS   (Fig. 93)

The radial face type seals used in the Turnbull split stern bearings are designed for replacement without the necessity of removing the shaft. The tailshaft can therefore be flanged at the inboard end for connection to the intermediate shafting and at the after end for connection to the propeller because all of the components of the seal can be split for replacement. Removal of the stern bearing allows a flanged shaft to be entered from aft.

The split seals can be examined or repaired by the methods described in the section on the Turnbull bearings. These seals are also used with conventional stern tubes arranged for oil or sea water lubrication.

Inboard and outboard seals are basically the same on oil lubricated stern tubes and each seal is adequate on its own while work is carried out on the other. Thus

113

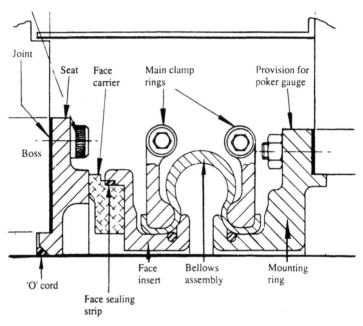

Fig. 93   Outboard stern shaft seal *(Manebrace)*

the inboard seal can be inspected and water entry is prevented by the outboard seal. The ship can be tipped for work on the outboard seal, leaving the inboard seal in place.

With the shaft rotating, the seals operate on a hydrodynamic film and are hydraulically balanced. The stationary face is attached to a bellows which is clear of the shaft and allows the face to follow any motion the shaft may make either from vibration or from differential expansion.

The seat of the outboard seal is bolted to the forward face of the propeller flange or to the propeller boss. It is jointed with asbestos fibre material and an 'O' ring is fitted for a taper mounted propeller. The rotating seat is a cast iron of high nickel content (about 14 per cent nickel) termed Ni-Resist which is able to stand up to the corrosive efects of sea water. The face seal is a synthetic termed ferrobestos. This is held in a carrier which, like the mounting ring, is of gunmetal. Clamp rings and the various nuts and bolts are of aluminium bronze. The mounting ring of the outboard seal is bolted to the stern frame with suitable jointing. The bellows assembly consists of monel metal springs in a cotton reinforced synthetic rubber. The seal is protected by a rope guard.

Materials of the inboard seal unit are not in contact with sea water so cast iron is used for the seat which is clamped to the shaft by a steel drive ring. The face carrier and mounting ring are of cast iron with steel clamp rings for the bellows. The face seal and bellows are of the same materials as used outboard but the springs in the bellows assembly may be of steel.

114

LUBRICATION SYSTEMS

The static lubrication system for vessels with moderate changes in draught have header tanks placed two or three metres above the maximum load waterline. The small differential pressure ensures that water is excluded. The cooling of simple stern tubes, necessitates keeping the aft peak water level at least one metre above the stern tube.

Tankers and other ships with large changes in draught, may be fitted with two oil header tanks (Fig. 94) for either fully loaded or ballast conditions. The sketch also shows a circulation and cooling system for the inboard seals which unlike those at the outboard end, cannot dissipate heat to the surrounding water. This circulation may be obtained by natural convection.

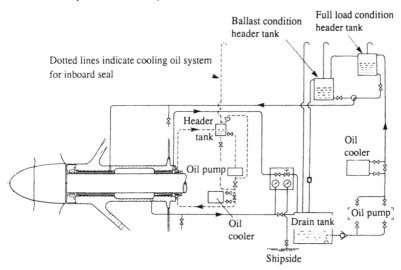

Fig. 94 Typical arrangement showing two header tanks for vessels with large changes in draught *(Glacier)*

HYDRODYNAMIC OR HYDROSTATIC LUBRICATION

The requirement for steaming at slow, economical speed during periods of high fuel price (or for other reasons) gives a lower fluid film or hydrodynamic pressure in stern tubes. The possibility of bearing damage, prompted installation of forced lubrication systems. The supplied oil pressure provides additional lift to separate shaft and bearing and a full flow for cooling.

SHAFT ALIGNMENT

Shaft systems would ideally be installed with straight alignment and remain in that state during ship operation. In practice there are many factors which affect and alter alignment during building and throughout the life of vessel.

The proposed centre line of the shafting system for a slipway built ship was used as the reference for accurate boring of the stern frame, prior to fitting of the stern tube with its propeller shaft and propeller. Propeller weight caused droop in the tailshaft and after launching the heavy and less buoyant stern section also flexed downward. Intermediate shafting aligned from the tailshaft, by paralleling of flanges, with no account being taken of tailshaft line and slight droop due to elasticity and overhanging weight at each shaft flange, would run out of line. Unless there was reference to a taut wire or other alignment aid, the engine would finally rest on chocks set lower than intended. Propeller weight and shaft sag, tends to impose edge loading on the aft end of the stern tube bearing. Slope boring reduces the edge loading problem.

The shaft line is continually changed through the lifetime of a ship, by hull flexure from different conditions of loading (cargo, ballast, fuel, fresh water). High deck and low sea temperature in the tropics cause differential expansion and hogging. Heavy weather produces changing conditions. The hull of a moderately sized ship can flex 150 mm in heavy weather. The local factors affecting shaft alignment include forward tilt of the thrust block, shaft lift as fluid film pressure builds up in bearings, sink of individual plumber blocks.

The method of fair curve alignment (developed at the Boston Navy Yard in 1954 and refined by others) accepts the changes of line endured by the shaft system and seeks a compromise to suit the varying conditions.

The initial calculation is to determine the load on each bearing, assuming all bearings to be in a straight line. The computer program then simulates the raising of each bearing through a range and calculates for each small change, the increase of its own load and alteration in load on each of the other bearings. The process is then repeated with a simulation of the lowering of each bearing in turn with the computer finding resultant load changes on the bearing in question and the others. Influence numbers, in terms of load change for each height variation, are calculated by the exercise, for all bearings.

The data bank of influence numbers enables the effects of changes in alignment from hull flexure and local factors to be found. All of the variables described for a 'pre-fair curve alignment' ship can be matched to find the best compromise for shaft installation.

SHAFT CHECKS

The intention of good alignment is to ensure that bearings are correctly loaded and that the shaft is not severely stressed. Alignment can be checked with conventionl methods, but results are uncertain unless the vessel is in the same condition with regard to loading and hull temperatures as when the shaft system was installed. Uneven bearing wear, hull deformation and other factors can affect results.

The method of jacking (Fig. 95) to assess correct bearing loads is used as a realistic means of ensuring that statically, the shaft installation is satisfactory.

The procedure involves the use of hydraulic jacks placed on each side of the bearing, to lift the shaft just clear. A dial gauge fixed to the bearing indicates lift. Hydraulic pressure, exerted by the jacks, registers the load on the bearing. A plot of lift and load is made.

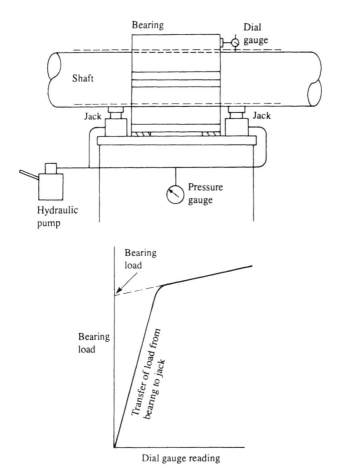

Fig. 95   Method of jacking to check bearing loads

References

Wilkin, T. A. and Strassheim, W. (1973). Some Theoretical and Practical
  Aspects of Shaft Alignment. *I. Mar. E.*, IMAS 73 Conference.

CHAPTER 8

Steering Gear

RUDDER CARRIER BEARING

The rudder carrier bearing takes the weight of the rudder on a grease lubricated thrust face. The rudder stock is located by the journal, also grease lubricated.

Support for the bearing is provided by a doubler plate and steel chock. The base of the carrier bearing is located by wedge type side chocks, welded to the deck stiffening. The carrier is of meehanite with a gunmetal thrust ring and bush. Carrier bearing components are split as necessary for removal or replacement. Screw down lubricators are fitted, and the grease used for lubrication is a water resistant type (calcium soap base with graphite).

An alternative type of carrier bearing with a conical seat (Fig. 97) has the advantage that the seat and side wall will locate the rudder stock. The angle of the conical seat is shallow to prevent binding.

Bearing weardown occurs over a period of time, and allowance is made in the construction of the steering gear for a small vertical drop of the rudder stock. Lifting of the rudder and stock by heavy weather is prevented by jumping stops between the upper surface of the rudder and the stern frame.

External rudder stops are fitted to limit its movement to, say, 39° each way from the mid position. In the steering gear there are also stops set to limit the

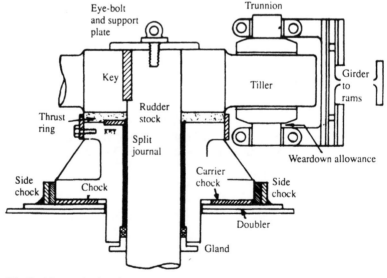

Fig. 96   Rudder carrier bearing

angle to which the rudder can be moved by the gear. These are set to, e.g., 37°
each way from the mid position. The latter are necessary to prevent the rudder
from being forced against the outside stops. The outside stops prevent unlimited
rudder movement, which could arise from damage, causing the rudder to
become disconnected. The result of this being, possibly, contact with the
propeller. Limits on the telemotor are set at 35° each way from the mid position.

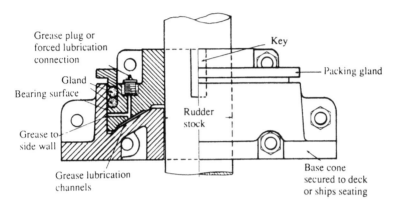

Fig. 97 Alternative carrier bearing *(Taylor-Pallister Dunstos type)*

RAM TYPE HYDRAULIC STEERING GEAR

The tiller, keyed to the rudder stock, is of forged or cast steel. For hydraulic
gears with four cylinders, the tiller has two arms, which are turned and ground
to slide in swivel blocks arrangement designed to convert linear movement of
the rams to the rotary movement of the tiller arms and rudder stock.

The rams are of close grained hard cast iron or steel with working surfaces
ground to a high finish. Each pair of rams is bolted together, the joined ends
being bored vertically and bushed to form top and bottom bearings for the
trunnion arms on the swivel block (see Fig. 96). Crosshead slippers bolted to the
sides of the rams, slide on the machined surfaces of the guide beam, so that the
glands in the cylinders are relieved of side loads. The guide beams also serve to
brace each pair of cylinders which tend to be pushed apart by the hydraulic
pressure on the arms. The cylinders have substantial feet bolted to the stools on
which the gear is mounted. In four cylinder sets adjacent cylinders are cross
braced by heavy brackets (not shown in the sketch of the four ram type) which in
conjunction with the guide beams, preserve alignment.

Hydraulic pressure is supplied by variable delivery pumps, with electric motor
drive, running at constant speed in the same direction. Pumps may be of the
Hele–Shaw radial piston type or of the axial piston V.S.G. In both, the stroke of
the pump pistons can be varied and flow of oil to and from the pump can be
reversed. When the operating rod of the pump is in mid position, there is no
flow of oil.

119

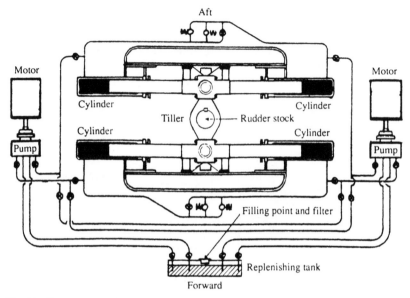

Fig. 98   Four ram steering gear diagrammatic arrangement

ROTARY VANE STEERING GEAR

The usual arrangement of three fixed and three moving vanes allows a rudder angle of 70° with a vane-type steering gear. A larger turning angle is obtained with two fixed and two moving vanes if required.

Vanes in the gear shown (Fig. 99) are of spheroidal graphite cast iron, the fixed ones being held to the stator by high-tensile steel dowel pins and cap screws. Moving vanes are keyed to the cast steel rotor which in turn is fitted to a taper on the rudder stock and keyed. Vanes are sealed by steel strips backed by synthetic rubber laid in slots. Weight of the gear is supported by a rudder carrier bearing beneath it in this design. Rotation of the gear is prevented by two anchor bolts held in fixed anchor brackets with rubber shock-absorbing sleeves. The bolts have outer cast-iron bushes to take wear from the steering gear flanges. Top and bottom stator flanges are welded on after oil manifold grooves have been machined.

PUMP NON-REVERSE LOCKING GEAR

The set can be run with both pumps in operation for quicker response although only one is normally used. When two pumping units are fitted and only one is running, the idle pump would be driven in the reverse direction by oil pressure from the other but non-reverse locking gear is fitted in the flexible coupling between the motor and pump. It consists of steel pawls on the motor coupling rim. These pawls with the motor and pump running, are thrown out towards the rim by centrifugal force. With the pump stopped, the pawls return to their

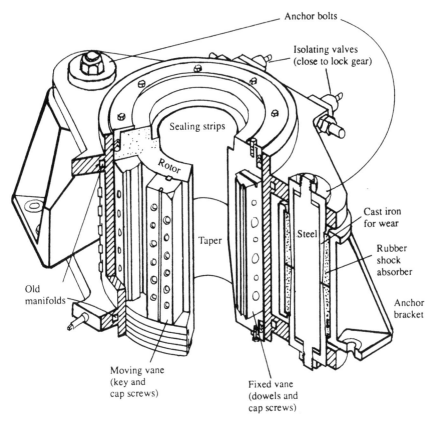

Fig. 99   Rotary vane steering gear

normal inward position and engage the teeth of a fixed ratchet secured to the pump base.

HELE–SHAW PUMP

The constant speed Hele–Shaw pump has its output controlled by a simple push/pull rod attached to guide rings in the pump. Without stopping or starting the pump, the output can be varied from zero to full in either direction.

The pump consists of a bronze cylinder body with seven or nine radial cylinders, which is rotated at constant speed in one direction (Fig. 100). The radial cylinder block rotates on a fixed steel central piece having two ports opposite to one another and in line with the bottom of the rotating cylinders. In each cylinder there is an oil hardened steel piston having a gudgeon pin with bronze slippers on the end. The slippers revolve with the cylinder block in grooves machined in a pair of floating rings. These are the rings which are moved horizontally by the control rod.

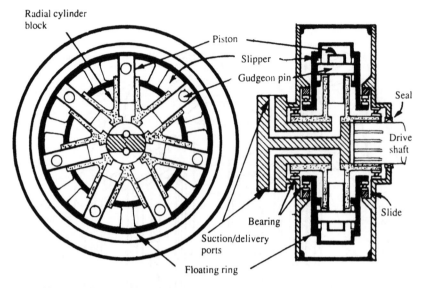

Fig. 100   Hele-Shaw variable delivery pump

Movement of the floating rings from the mid position displaces the circular path of rotation of the pistons from that of the cylinder block and produces a pumping action. When the rod is in mid position and the centres of rotation of pistons and block coincide, there is no pumping action.

V.S.G. PUMP

The V.S.G. pump has a cylinder block with axial cylinders. The sketch (Fig. 101) shows a simplified arrangement. Piston stroke and oil flow are varied by angular movement of the swash plate.

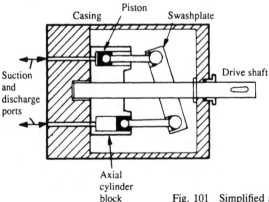

Fig. 101   Simplified arrangement of V.S.G. pump

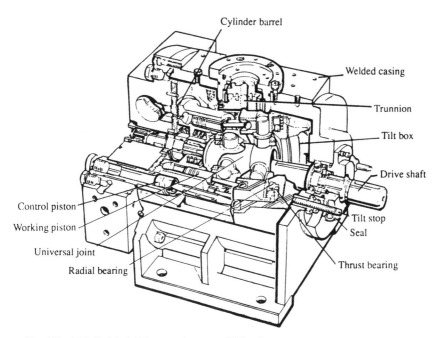

Fig. 102   V.S.G. Mark IV pump *(courtesy Vickers)*

The V.S.G. pump has been developed in recent years to operate with higher pressure with a resulting decrease in size of steering gears. The unit shown (Fig. 102) accordingly has a casing of fabricated steel rather than cast iron. The cylinder block with its pistons is driven through the drive shaft by a simple constant-speed electric motor. The pistons are tied through piston rods and bearings to a swash plate or tilting box. With the latter vertical, the pistons rotate with the cylinder block but have no axial movement. When the swash plate or tilting box is set at an angle by the controller the pistons are caused to reciprocate in their cylinders and produce a pumping action. Stepless changes of pump delivery from zero to maximum in either direction is achieved, through lever or servo controls.

OPERATION OF FOUR RAM GEAR

The pipe arrangement (Fig. 98) on the simple sketch of the four ram system shows the connections from two pumps to four rams. Because of the non-reverse lock arrangement, one pump can be stopped with the valves left open. All four rams work together but in the event of damage, either pair of rams can be isolated from the pressure pumps and allowed to idle with the bypass open. The remaining pair of rams then operate as a two ram gear. The by-passes are in parallel with the relief valves.

The rudder can be locked by closing the supply valves, in an emergency.

The relief valves between pipes connecting the opposing rams are designed to lift if pressure in the system rises to about 10 per cent above normal. This will occur due either to the rudder being hit by a heavy sea or from direct loading. By-passing of oil from one side of the system to the other through the relief valves permits the rams to move and abnormal stress on the rudder stock is thereby avoided. The hunting gear will cause the rudder movement to be corrected by putting the pump on stroke.

Each pump has suction connections through non-return valves from the replenishing tank. Losses of oil from the system are automatically made up from this reserve of oil. A certain amount of leakage occurs in the pump and this oil is drained to the replenishing tank. Where an overhead tank is fitted (V.S.G. pump) the oil is caused to flow from the pump casing by a centrifugal action produced by rotation of the cylinder block. Other oil flowing back into the pump gives a cooling action.

HUNTING GEAR

The pump control is moved by the telemotor through a floating lever. The other end of this lever, is connected through a safety spring link to the rudder stock or tiller (Fig. 103).

The telemotor is the receiver of the hydraulic remote control system from the wheel on the bridge. The linkage through the floating lever of telemotor, pump and rudder stock forms the hunting gear.

The pump is only required to deliver oil when the steering wheel is moved. The hunting gear returns the pump operating rod to mid position as soon as the helmsman stops turning the wheel. When the rudder has moved through the angle corresponding to the wheel position, it will remain there until the wheel and telemotor are moved again.

The sketch shows simply, the operation of the hunting gear. The telemotor moves the end of the floating rod A to A_1 and the pump control is moved, therefore, from B to B_1. Pumping of the hydraulic oil causes movement of the rams and the end of rod C moves to C_1 thus causing the pump control to be pulled back to the neutral position B.

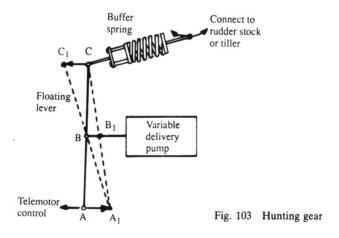

Fig. 103   Hunting gear

If the rudder is displaced by a heavy sea through lifting of the relief valves, the hunting gear is moved by the rudder stock. This will put the pump on stroke and the rudder will be restored to its previous position.

TELEMOTOR

The telemotor has become, on many vessels, the standby steering mechanism, used only when the automatic steering fails. It comprises a transmitter on the bridge and a receiver connected to the variable delivery pump through the hunting gear. Transmitter and receiver are connected by solid drawn copper pipes. Liquid displaced in the transmitter causes a corresponding displacement in the receiver and of the pump control.

The transmitter (Fig. 104) consists of a cylinder with a pedestal base which contains a piston operated by a rack and pinion from the steering wheel. The make up tank functions automatically through spring loaded relief and make up valves. Excess pressure in the telemotor system causes oil to be released through the relief valve to the make up tank and loss of oil is made up through the lightly loaded make up valve. The two valves are connected to the cylinder through a shut off valve (which is normally left open) and the by-pass which connects both sides of the pressure system, when the piston is in mid position. There is also a hand operated by-pass.

The tank must be kept topped up. The working fluid is a mineral oil of low viscosity and pour point and this gives some protection against rusting. As an alternative to mineral oil, a mixture of glycerol (glycerine) and water has been used.

The receiver in the steering gear is shown as two fixed rams with moving opposed cylinders. Centering springs are fitted to bring the cylinders to mid position. Movement of the telemotor receiver is limited by the stops set at 35°.

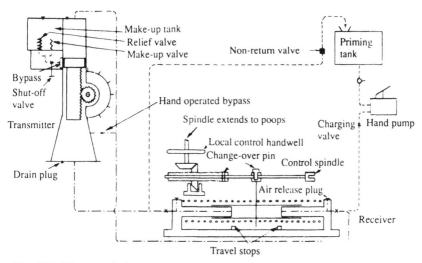

Fig. 104   Diagram of telemotor system

The telemotor is connected to the hunting gear through a control spindle. By switching the change-over pin, the control spindle can be operated from the emergency steering position on the poop.

CHARGING THE SYSTEM

The system is provided with a priming tank and hand pump which are situated in the steering gear compartment. When charging, the shut off between the make up tank and transmitter is closed and the wheel is brought to the mid position so that the piston is at the centre of its travel and the top and bottom parts of the cylinder are connected through the bypass. The priming tank is filled (and then kept topped up as necessary) and the hand pump operated with the charging valve open. Each section of the pipe is progressively filled with air being released through the bleed screws. At the last section the non-return valve is opened to allow oil from the end of the pipe to be returned to the priming tank. Pumping is continued for some time, then the non-return valve is closed while pressure is maintained with the pump. The bleed screw at the top of the cylinder is cracked open to get rid of any remaining air. After closing the bleed screw on the cylinder, the shut off between the cylinder and make up tank is opened and the tank is brought up to level. For an initial charge, pipes are disconnected so that the sections of pipe can be washed through.

During the charging operation, joints are checked for leakage and when the system is full, a further check is made with the shut off closed and pressure maintained with the pump. At this stage, if the non-return valve is opened, each stroke of the hand pump should produce a discharge back to the tank which exactly coincides with the movement of the pump lever.

The system is made ready for testing and operation by closing the charging and non-return valves and opening the shut-off valve. The hand by-pass valve must be closed.

CHARGING STEERING GEAR

The steering gear itself must be completely filled with oil and all air must be excluded. Thus the air release valves are opened on hydraulic cylinders and pumps, also stop valves and by-pass valves in the system. The variable delivery pump can be used to pump the oil around the system (while keeping the replenishing tank topped up). It can be put just on stroke by the handwheel (Fig. 104) and turned by a bar. The rams may be filled through the filling holes until all air has been displaced, before starting to pump the system through.

When all air has been purged from the system and the level in the replenishing tank ceases to fall, the air release valves are closed. Finally the by-pass and stop valves are set for normal running and the pump is started. Using the hand control, the gear is then run from hard over to hard over slowly and the air release valves are again checked.

Charging methods for steering gears and telemotors vary from one type to another.

STEERING GEAR CHECKS

Pollution resulting from tanker casualties, where the blame was attributed at least in part to failure of the steering gear or its control system, drew attention to the possible need for changes in international regulations as applied to steering equipment. The focus of attention was on tanker steering gears, but there are additional requirements now for other ships as well.

The test procedure to be carried out not more than 12 hours before departure (or weekly on short-voyage vessels) requires operation of (where applicable) the following: main steering gear, auxiliary gear, remote control systems, bridge steering position, emergency power supply, rudder indicator, power failure alarms for remote control systems and also for the steering gear power unit. While the gear is running, full movement must be checked and a complete visual inspection must be carried out. The phone or other means of communication between bridge and steering compartment must be tested also.

While the ship is at sea with the automatic pilot in prolonged use, manual steering must be tested before entering busy or restricted waters. When in such waters both power units (pumps and motors) must be running if simultaneous operation is possible.

At three-monthly intervals emergency drills (including local control and communication; and operation with alternative power) should be carried out. All officers are required to be familiar with the steering gear and the changeover arrangements. Instructions for changeover must be displayed in the steering compartment and on the bridge. The various tests and checks should be logged.

STEERING GEAR FAILURES AND SAFEGUARDS

Equipment installed to operate the rudder can be duplicated, but conventional ships have only one rudder and rudder stock. These two items and any others not duplicated must therefore be of a strength sufficient to make failure unlikely.

Failure of the main steering gear has been safeguarded against by provision of an emergency gear or duplication of the power units (e.g. two pumps fitted with the one main gear). The steering gear itself is made of substantial strength and often duplicated, as where a four-ram instead of two-ram type is fitted. However, hydraulic steering gears whether of the two or four ram (or double acting) type, or of the single or double chamber vane type, can be rendered useless by loss of oil from the hydraulic system. The loss, probably resulting from a fractured pipe or failure of flange studs, would be most likely to occur when the pump was on stroke and system pressure high. In the *Amoco Cadiz* it was also thought that an additional surge in system pressure was produced by a heavy sea striking against the rudder, in the opposite direction to that in which it was moving. Pipe/flange failure at such a time with the pump perhaps on full stroke would mean that all of the hydraulic oil would be discharged into the steering compartment in a matter of seconds. In bad weather conditions, the effect of sea and ship movement would be to make rudder and steering gear swing wildly from side to side with the final result of a smashed gear. This has happened in a few incidents with efforts of personnel to rectify the situation being hampered by oil on the deck. Arresting the movement with a brake is not possible because the only brake available is that provided by closing valves on the hydraulic system—not effective with no oil in the circuit or cylinders. Very few ships have been fitted with friction brakes. In one episode, coils of rope were thrown in to finally jam a gear which had lost oil through a fractured pipe, but too late to prevent damage beyond repair. It is probable that movement of the rudder and steering gear could be cut down by keeping the propeller turning to maintain forward movement of the ship through the water and therefore a slipstream over the rudder.

Statistics show a decrease in hydraulic system reliability with increase in system pressures and age. (High pressures of 170 bar permit small equipment size.) The failures might be due to surges being more extreme in the contemporary high-pressure plant. Discussion at international level, when considering the single failure concept in relation to the hydraulic system, the major pollution risk with loss of steering on a large tanker, and the statistics of steering gear failure, resolved that it will be necessary to have complete hydraulic system

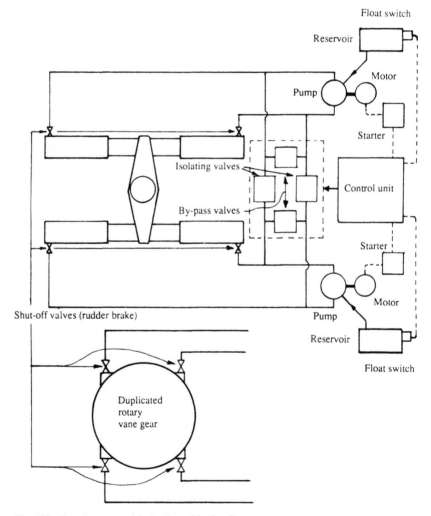

Fig. 105   Steering gear with duplicated hydraulic system

redundancy on new tankers of 100,000 dwt and above. This implies either two complete and independent hydraulic steering gears or two interconnected circuits with automatic isolation of one from the other should there be a loss of hydraulic fluid (Fig. 105).

INTERNATIONAL REGULATIONS

These now require either that cargo vessels in general must have an auxiliary steering gear as a backup for the main gear (and that if the rudder stock is over 356 mm (14 in.) diameter this second gear must be power operated) or that power units and connections are duplicated. 'Power unit' describes the pump and motor or equivalent. Short-circuit protection only is to be provided for electric motors and power circuits. There must be two widely separated power circuits from the main switchboard, one of which may pass through the emergency switchboard. A rudder indicator must be fitted on the bridge.

PASSENGER VESSELS   At maximum service speed the main steering gear of a passenger ship must be able to move the rudder from 35° on one side to 35° on the other, and 28 seconds is allowed for that part of the movement from 35° to 30°. Main and auxiliary gears are required or duplicated pump and motor power units. Any auxiliary steering gear must be power operated if the rudder stock is over 230 mm (9 in.). If the duplicated power unit alternative is used, each pump and motor (or equivalent) must be capable of meeting the performance criteria.

TANKERS   Stricter international regulations for the steering gears of tankers over 10,000 tons gross were made effective for new vessels from 1980 and at a later date for existing ships.

The steering gear must be able to meet the performance requirement of being able to move the rudder from 35° on one side to 35° on the other with 28 seconds being allowd for that part of the movement from the 35° extreme to 30° at the other. Main and auxiliary gears are required or there may be duplication of power units. Design of the system should be such that a single failure in its piping or one power unit (pump and motor or equivalent) will not leave the steering gear inoperable.

Duplicated and widely separated electrical supply circuits are required from the main switchboard with short-circuit protection only for these and the motors. Failure alarms are to be fitted on the bridge, with manual or automatic means of restarting the power unit motors.

As well as the two bridge steering gear controls with audible and visual alarms, a local control in the steering gear compartment is also required. Rudder position is to be indicated in the steering gear compartment as well as on the bridge and means of communication provided.

Main steering gear power units must be arranged to restart automatically when the electrical supply is restored after a failure. The possibility of total loss of electrical power is to be guarded against, by provision of alternative power for operation of the steering gear, the bridge control and the rudder indicator. The emergency supply is to be automatically connected within 45 seconds of main supply failure and must be capable of continuous operation for 30 minutes. The standard of performance when the equipment is working on the alternative supply is that at least it will move the rudder from 15° on one side to 15° on the other in 60 seconds. It must be capable of this with the ship at its deepest draught and running ahead at one-half the maximum ahead service speed or 7 knots, whichever is greater.

The alternative power supply can be taken from the emergency source of electrical power. Other suggestions are for an independent power source used solely for steering and located in the steering compartment, such as batteries, diesel or air motor pump drive. Air motors have been fitted in a number of ships.

It should be noted that Government and Classification Society regulations while broadly similar to the IMO rules outlined above, differ in some repects and may be more stringent.

STEERING GEAR FOR LARGE TANKERS AND OTHER VESSELS

The principle of a steering gear suitable for any vessel, including tankers of more than 100,000 dwt, is shown in Fig. 105. The four-ram gear consists of two pairs of rams, each pair being capable of supplying 50% of the torque required. For normal full ahead running they are operated together to provide 100% torque, with one pump and motor power unit in use. The system operates in the same way as other four-ram arrangements, but duplication of the hydraulic pipework as well as pump and motor power units gives an additional safeguard with complete hydraulic system redundancy.

The same sort of design can also be used in conjunction with duplicated rotary vane cylinders (one chamber above the other) as indicated in the sketch.

Oil loss from a fracture in the pipe system would lower the level in the reservoir of the running pump and through the float switch and control unit shut down the isolating valves. The two sets of piping and associated pairs of rams would now be isolated. The leak could be in either pipe and ram set, however, and the problem remains whether to shut down the running pump and start the other. One proposal is that a second lower level float switch be fitted to each reservoir. If, after closure of the isolating valves, no further oil loss occurred, the running pump would be left in operation, but continuing drop in oil level would initiate shutdown of the running pump and start of the other in 45 seconds.

A number of proposals have been advanced for shutdown and isolating arrangements. There is the risk that the apparent increase in safety is jeopardised by added complexity and greater number of components exposed to failure.

References

Cowley, J. (1982). Steering Gear: New Concepts and Requirements. *Trans. I. Mar. E.*, vol. 94, paper 23.

The Merchant Shipping (Passenger Ship Construction and Survey) Regulations 1984. HMSO.

The Merchant Shipping (Cargo Ships Construction and Survey) Regulations 1984. HMSO.

CHAPTER 9

Pollution Prevention— Monitoring—Oily Water Separator—Sewage Treatment

OIL POLLUTION PREVENTION

A major source of oil pollution in the past from the operation of ships was the discharge into the sea of tank washings from tankers. This was reduced by the discharge of tank washings to a slop tank for settling, and discharge overboard of the water while retaining the sludge for pumping ashore to the refinery, with the next cargo. Crude Oil Washing (COW) eliminates the use of water and enables cargo residues to be pumped ashore during cargo discharge because cleaning is carried out simultaneously with the discharge.

Ballast carried in oil cargo and bunker tanks which is therefore contaminated with oil constitutes another pollution source, unless pumped out via an oily water separator. New regulations require tankers of certain sizes to have segregated or clean ballast tanks.

A third pollution source is from machinery space bilges.

Previous legislation intended to prevent pollution has been superseded by new rules that came into force in October 1983 and which are set out fully in the IMO publication 'Regulations for the Prevention of Pollution by Oil'.

In brief, if oil cargo residue is to be discharged from a tanker it MUST: (a) not be in a special area, (b) be farther than 50 nautical miles from land, and (c) be on passage. Additionally (d) the instantaneous rate of discharge must not be more than 60 litres per mile, and (e) the total amount must be not more than 1/30,000 of the particular cargo (previously the amount was 1/15,000 of the cargo, and this still applies to older vessels). The Oil Record Book for cargo/ballast operations which is kept by the deck department on all tankers over 150 gross tonnage is now to have chronological entries with date, operational code and item number in appropriate columns. Items and codes are listed on the first page of the book. An Oil Record Book for cargo/ballast operations is also required on any vessel with a bulk oil capacity in excess of 200 m^3.

Engine room bilge disposal is treated separately in the regulations from discharges related to cargo/ballast operations. The bilges must only be pumped through suitable oily water processing equipment or retained for discharge ashore, if due to the small size of the ship or other reason suitable equipment is not fitted. The oil content of any discharge must be less than 100 p.p.m. In special areas and within 12 miles of land, it must be less than 15 p.p.m.

In general, ships over 400 g.t. are permitted to discharge machinery space bilges into the sea provided: (a) the oil in the bilge discharge does not amount to more than 100 p.p.m., (b) this is achieved by the operation of an oily water separator and/or filtering system with discharge monitoring and control, (c) the ship is moving on passage, (d) it is at least 12 nautical miles from land, and (e) is not within a Special Area.

In Special Areas (Mediterranean, Baltic, Black Sea, Red Sea, Persian Gulf area) bilge discharge is permitted only when an oily water separator or filtering system capable of reducing the oil content to below 15 p.p.m. is fitted and in use. Bilge discharge monitoring and control equipment with alarm and automatic stopping device must also be fitted and in use. Discharge within the 12 nautical mile limit is allowed under the same conditions as for Special Areas except that the alarm and stopping device are not mandatory.

Regulations applied to vessels of less than 400 tons gross are not as stringent. Rules for ships of over 10,000 g.t. and some others are more strict.

The Oil Record Book for machinery space operations, which must be kept by all ships of 400 tons and above, is the responsibility of the engine room department. Items to be logged are: (a) oil fuel tank ballasting and cleaning, (b) discharge of the dirty ballast or cleaning water, (c) sludge/oil residue disposal, (d) discharge or disposal of bilges, (e) automatic bilge discharge, (f) oil discharge monitoring/control system failure details, (g) oil discharges (accidental or exceptional).

OILY WATER SEPARATOR

The performance of separators has been improved to meet the requirements of stricter regulations by the addition in some designs of a second stage coalescer.

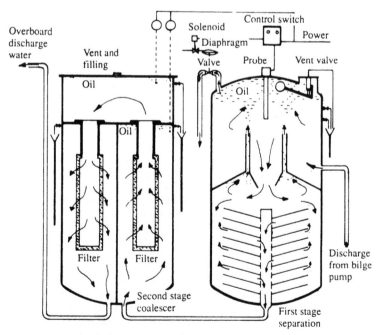

Fig. 106   Simplex Turbulo oily water separator with coalescing second stage

Filter elements in the second stage remove any small droplets of oil in the discharge and cause them to be held until they form larger droplets (coalesce) which rise to the oil collecting space.

The first stage of the Turbulo separator is supplied with oily water from the pump. Because of the different densities, the water and oil will start to separate in the top part of the chamber, the oil tending to rise to the upper part. Further separation occurs in the lower chamber where the liquid has to pass through a series of dished plates before leaving the first stage separator. Oil droplets from the plates tend to travel upwards, being collected at the baffles and funnelled up to the oil collection space. Oil from the top of the chamber is automatically drained to the oil tank. Air is vented by the float controlled release valve.

The oil drain valve from the top of the first stage separator is a diaphragm controlled piston valve. Access of control air to the diaphragm is through a solenoid operated pilot valve. The capacitance probe senses oil quantity in the collection space and causes the solenoid to be energized through the control switch.

Water from the first stage passes downwards through a central pipe to the second stage coalescer. The filter in the right hand chamber removes solids and some oil. Coalescing inserts in the left hand chamber take out the remainder of the oil in the form of small droplets which coalesce to form larger drops. These rise to the oil collecting space. Oil from the second stage coalescer is drained manually, at intervals, through the cocks provided.

Oil content of the final discharge is below 15 p.p.m.

OIL CONTENT MONITORING

Inspection glasses fitted in the overboard discharge pipes of engine room oily water separators allowed sighting of the flow. The discharge was illuminated by a lightbulb fitted on the outside of the glass port opposite the viewer. The separator was shut down if there was any evidence of oil, but problems with observation occurred due to poor light and accumulation of oily deposit on the inside of the glasses.

Present-day monitors are based on the same principle. However, while the eye can register anything from an emulsion to globules of oil a light/photo-cell detector cannot. Makers may therefore use a sampling and mixing pump to draw a representative sample with a general opaqueness more easily registered by the simple photo-cell monitor. Flow through the sampling chamber is made rapid to keep the glasses clean and they are easily removable for cleaning.

Bilge or ballast water passing through a sample chamber can be monitored by a strong light shining directly through onto a photo-cell (Fig. 107). Light reaching the cell decreases with increasing oil content of the water. The effect of this light on the photo-cell compared with that of direct light on the reference cell to the left of the bulb can be registered on a meter calibrated to show oil content.

Another approach is to register light scattered by oil particles dispersed in the water by the sampling pumps (Fig. 108). This light when compared with the source light increases to a maximum and then decreases with increasing oil content of the water. Fibre optic tubes are used in the device shown to convey light from the source and from the scattered light window to the photo-cell. The motor-driven rotating disc with its slot lets each light shine alternately on the

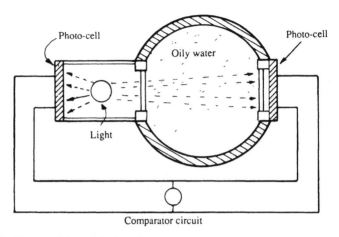

Fig. 107 Director light monitoring chamber

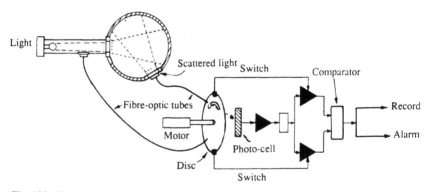

Fig. 108 Sofrance type scattered light monitoring chamber

photo-cell and also by means of switches at the periphery causes the signals to be passed independently to a comparator device.

These two methods briefly described could be used together to improve accuracy, but they will not distinguish between oil and other particles in the water. Methods of checking for oil by chemical test give truer results but they take too long in a situation where excess amounts require immediate shutdown of the oily water separator.

TANKER BALLAST Sampling and monitoring equipment fitted in the pump room of a tanker can be made safe by using fibre optics to transmit light to and from the sampling chamber (Fig. 109), the light source and photo-cell being in the cargo control room together with the control, recording and alarm console. The sampling pump can be fitted in the pump room (to keep the sampling pipe

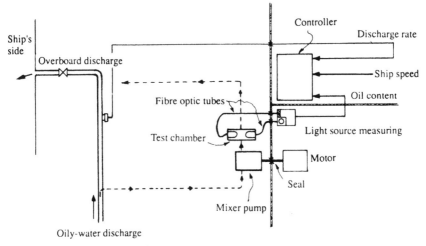

Fig. 109   Seres type oil monitoring system

short and minimise time delay) with its drive motor in the machinery space and shaft passing through a gas-tight seal in the bulkhead.

Oil content reading of the discharge is fed into the control computer together with discharge rate and ship's speed to give a permanent record of the monitoring. Alarms, automatic shutdown, back-flushing and recalibration are also incorporated.

SEWAGE

Raw sewage discharged in restricted waters will eventually overwhelm the self purification ability of the limited quantity of water. In a closed dock the effects of pollution can be clearly seen from the black sludgy water which, when stirred by the ship's propeller, gives off an unpleasant smell.

Where the amount of sewage relative to water is small, dissolved oxygen in the water will assist a bio-chemical (aerobic) action which breaks down the sewage into simple components and carbon dioxide. In the natural cycle, these in turn help to produce plant life which returns oxygen to the water.

Decomposition occurs both with excessive quantities of waste and with small amounts. In the former the decomposition is termed anaerobic and is associated with the production of unpleasant odours and putrefaction. While it is the principle of operation of the septic tank, it is not suitable for sewage treatment on a ship. Ship sewage treatment can be carried out using the aerobic breakdown process and the supply of oxygen is maintained by bubbling air through the water. Oxygen in the air promotes the multiplication of the bacteria and satisfactory decomposition of the wastes. The bacteria build up a colony using the sewage as food and oxygen for their metabolism. The action results in the production of a clean effluent liquid which is disinfected and discharged overboard.

There are alternatives to the aerobic sewage treatment plant such as physical/

135

chemical treatment systems in which breakdown is achieved with the aid of chemicals, and a circulation process with sludge removed at intervals.

AEROBIC TREATMENT PLANT
(Flow-through System)

The Trident sewage treatment system shown (Fig. 110) has four compartments. Incoming waste passes through a coarse screen into the primary collection tank where it remains until displaced by overflow into the aeration section. A connection is provided so that the primary collection tank can be pumped out.

Breakdown of the wastes in the aeration compartment is induced by bacteriological organisms promoted by the presence of oxygen. The oxygen is supplied by the air from the blowers which enters the aeration section through a fine bubble diffuser at the bottom. The diffuser is of porous material so that clean air is needed to prevent blockage. The bubbles, besides providing oxygen, also create turbulence so that settlement is prevented and good mixing obtained.

After prolonged aeration, the mixed liquor is displaced into the settlement tank where the biological floc is formed. Activated sludge gravitates to the bottom and is continuously withdrawn and returned to the aeration chamber, to mix with incoming wastes. In the unit shown, the sludge is returned to the aeration compartment by an air lift. Clean effluent from the top of the settling tank is collected in the last compartment for disinfection and discharge overboard.

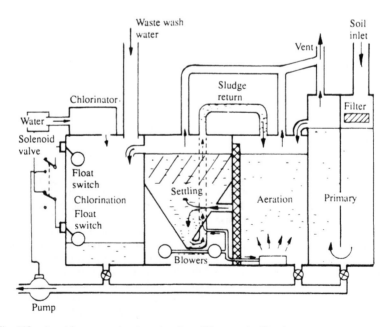

Fig. 110   Aerobic sewage treatment system *(Hamworthy Type)*

Two float switches in the final section control the discharge pump. As the tank fills, the top float switch starts it. When the low level float switch stops the pump, it simultaneously opens a solenoid valve in the water supply line to the hypochlorinator. In passing through the chlorinator the water forms a sterilizing solution. A timing device on the solenoid valve allows the correct amount of solution into the treatment tank to deal with the next charge. There are various means of sterilization and tablets of compound used in one method, may become unstable and dangerous during storage.

The bacteria in the aerobic treatment plant must be kept alive by maintaining the correct conditions. They are sensitive to temperature, type of water and regularity of flow. If the installation is shut down for a long period, the bacteria die so that a new colony has to be started.

Regular removal of sludge will prevent impairment of operation. The sludge is pumped overboard when the vessel is at sea clear of areas where restrictions are in force. Small quantities can be burnt in an incinerator. These are fitted in some vessels. Effluent discharged overboard through the sewage plant when the vessel is in port or restricted waters, will have to meet certain standards. Thus tests have been established. One is simply a check of the quantity of solid material in the effluent; two others involve incubation of samples and would probably be carried out in a laboratory.

SUSPENDED SOLIDS

In this test, the quantity of solid material in the effluent, is checked by collection and weighing. An asbestos mat filter element is used and the solids are collected on this then dried and weighed. Test results are in parts per million or milligrams of suspended solids per litre.

BIOCHEMICAL OXYGEN DEMAND

Bacteria decompose sewage and in the process use oxygen. At the end of the process the sewage is said to be stable and as the activity of the bacteria reduces so the oxygen consumption also drops. The effectiveness of sewage treatment can be gauged by taking a sample (one litre) and incubating it for five days at 20°C with well oxygenated water. The amount of oxygen taken up by the sample in milligrams per litre or p.p.m., is termed the Biochemical Oxygen Demand (B.O.D.)

COLIFORM COUNT

There are certain tell tale bacteria found in human waste which originate from the intestine. These are coliform organisms. Disinfection of the effluent at the end of the sewage treatment process will reduce the coliform level and also the level of other organisms which may be present such as those responsible for typhoid, dysentry, gastro-enteritis etc. Apart from these bacteria found in raw sewage there are also viruses of the type responsible for illnesses like poliomyelitis and infectious hepatitis.

The effectiveness of disinfection is checked by a coliform count, carried out on a sample of effluent. Results are given as the number of coliforms per 100 ml of effluent. One coliform test consists of incubating a sample over a 48 hour period at 35°C. Another test takes 24 hours to produce a colony of bacteria at an incubation temperature of 35°C.

REGULATIONS

Legislation to prevent the discharge of raw sewage into docks, harbours and other closed waters of certain countries has been in existence for some time. Deadlines have been set for the installation of sewage treatment plant on vessels trading to American ports.

American legislation defines three types of sewage treatment plant:

Type 1—A device capable of discharging effluent having no visible floating solids and a coliform count of less than 1000 per 100 ml of effluent.

Type II—A device capable of discharging effluent with suspended solids not in excess of 150 mg/litre and a colliform count of less than 200 per 100 ml.

Type III—A device to prevent the discharge overboard of treated or untreated sewage.

ELSAN (ZERO DISCHARGE) SYSTEM

A retention or holding tank is required where no discharge of treated or untreated sewage is allowed in the ports of certain countries. The sewage is then pumped out to shore reception facilities or overboard when the vessel is again proceeding out at sea (outside a 12 nautical mile limit).

Straight holding tanks for retention of sewage during the period of a ship's stay in port were of a size large enough to contain not only the actual sewage but also the flushing water. Each flush delivered perhaps 5 litres of sea water. Passenger vessels or ferries with automatic flushing for urinals required very large holding tanks.

Other problems resulting from the retention of untreated wastes related to its breakdown by anaerobic bacteria. Clean breakdown by aerobic organisms occurs where there is ample oxygen as described previously. In the stagnant conditions of a plain retention tank where there is no oxygen, anaerobic bacteria and other organisms thrive. These cause putrefaction, with corrosion in the tank and production of toxic and flammable gases.

The Elsan type system (Fig. 111) incorporates chemical treatment of the sewage to be retained. A perforated rubber belt is used to separate liquid from solids in the separating tank. The liquid is then passed through treatment tanks to a pneupress arrangement for use as a flushing fluid. Treatment by chlorine and caustic based compounds makes the liquid effluent acceptable for the purpose.

Solids are chemically inerted by a caustic compound and delivered via grinder pump to the holding tank. Capacity of the tank is 2 litres per person per day. The tank is pumped out at sea, or to shore if the ship is in port for a long period. Tank size is small because liquid effluent passes mainly to the flushing system (excess overflows to the sullage tanks).

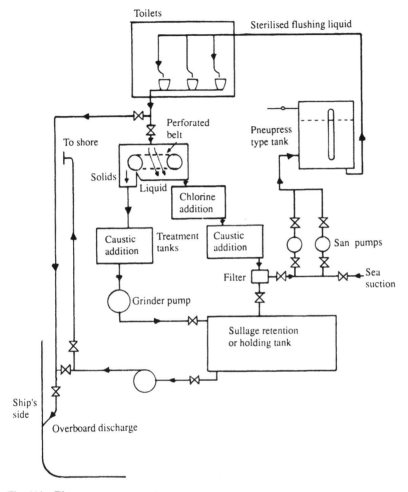

Fig. 111   Elsan type sewage system

SEWAGE REGULATIONS

Annex IV of Marpol 73/78 (IMO) has the aim of regulating the disposal of sewage from ships, internationally. Certain countries had already anticipated any internationally applied rules with their own national and regional controls.

In general, untreated sewage should only be discharged outside the 12 mile limit; comminuted and disinfected sewage could be discharged as near as 4 miles from land; nothing should be discharged within the 4 mile limit. The ruling, as applied to vessels of over 200 g.t. with more than ten personnel, would give a ten-year grace to existing vessels. Many countries dump their own sewage practically on the shoreline.

AIR POLLUTION

The Clean Air Act in Britain is intended to discourage emission of black smoke from the chimneys of industrial and other premises and from ships. Other countries have their own equivalent regulations. Failure to comply with them may lead to a court case and a heavy fine.

Smoke from the funnels of vessels in docks or harbours is checked for 'blackness' by comparison with the Ringelman Chart. If darker than shade 4 it is considered to be over the limit and to be black smoke. In other countries the criterion may be different (e.g. black smoke visible through a darkened glass). Obviously, black smoke is sometimes inevitable when flashing up or when a problem occurs. Regulations therefore usually include a specific time limit during which black smoke emission is not penalised: in the U.K. its continuous production must not be for longer than four minutes. Usually the combustion system would be shut down again or adjusted fairly quickly at emission and in a shorter time than four minutes. Thus there is also a regulation applied to repeated attempts to flash up and reduce smoke, where there are a number of short emissions: the aggregate time must be limited to three minutes in any twenty minute period. Another part of the ruling limits emission to not more than ten minutes in the aggregate in any two hour period.

Part of the Clean Air Act is taken up with examples of the line of defense that may be taken in court. This acknowledges that there are occasions when due to problems with combustion equipment or fuel, black smoke emission may be unavoidable.

The overall responsibility of the Chief Engineer and others in his department is to ensure by maintenance, care, training and watchkeeping (or checking of control equipment and alarms) that no smoke is produced in closed waters or, for the sake of efficiency, at any time.

PREVENTION OF POLLUTION FROM BULK CHEMICAL CARRIERS

Until the regulations to control and prevent pollution of the sea by chemicals came into force on the 6th April 1987, there was no real restriction against discharge of cargo remains or tank washings from whatever cargo remained in the tanks of chemical tankers. The only factors limiting pollution were goodwill and the fact that cargo remaining in tanks after discharge constituted a loss to the shipper. For some tankers there were substantial remains because of the inability of the older type of cargo pumps to discharge completely. Later generations of cargo pumps (see Chapter 2) were designed for more efficient discharge so that cargo tank remains were minimal. Improved clearing of tanks anticipated ideas put forward in draft regulations for more complete discharge of cargo, as a means to reduce pollution. Special draining and discharge methods have also been produced for fitting to existing vessels.

Regulations which came in force on the 6th April 1987 divide bulk liquid chemical cargoes into four categories (A, B, C and D) and give general directions for discharge and tank washing. There is a requirement in the rules for a Cargo Record Book, and a Procedures and Arrangements Manual to be carried as a reference.

The list of *Type A* chemicals includes: acetone cyanohydrin; carbon disulphide; and cobalt naphthenate in solvent naphtha.

The discharge into the sea of type A substances, and any initial washings which carry them, is prohibited. Chemicals in this category have to be totally discharged and delivered to the shore. Thus, when discharge is complete any cargo remains must be removed and also discharged ashore by washing through. The washing process is continued until the content of the type A chemical falls below a certain value. After this, the discharge from the tank must continue until the tank is empty.

The washing through to clear cargo is solely for that purpose and not intended as a complete cleaning operation. Traces of type A cargo on the surfaces of tank bulkheads will remain until removed by a subsequent washing operation. These washings are considered as forming a residual mixture constituting a hazard if freely discharged. The rules are extended to include those for disposal of the subsequent tank washing operation residue.

Only wash water added after cargo discharge and completion of the 'in port' washing routine can be pumped overboard at sea. A ship's speed being not less than seven knots, with the vessel more than 12 nautical miles from land and in water depth of 25 metres minimum, the effluent may be pumped out through a discharge situated below the waterline and away from sea inlets.

A special low capacity pump which leaves the offending liquid mix in the film of water flowing over and adjacent to the hull is used. It is intended that this flow shall carry the chemical into the propeller where it will be broken up and dispersed in the wake. Presence of chemical after this would not, in theory, exceed 1 p.p.m.

For most type A chemicals the content in the pre-wash to the shore must be reduced to less than 0.1 per cent (weight) while in port, if later washings are to be discharged outside special areas. If discharge of the later washings is to be in special areas (Baltic and Black Sea etc.) then port washing must, in general, reduce content of category A chemicals to less than 0.05 per cent by weight. Carbon disulphide is an exception for which content must be less than 0.01 (not in special areas) and 0.005 (special areas).

Type B chemicals include: acrylonitrile; some alcohols; calcium hypochlorite solution; and carbon tetrachloride.

The cargo pumps for the type B substances in chemical tankers built after 1st July 1986 must be capable under test, of clearing 'water' from the tank such that remains do not exceed 0.1 cubic metres (0.3 cubic metres for older vessels). Guidance for discharge ashore of category B cargoes is obtained from the Procedures and Arrangements manual. Where difficulties prevent discharge according to the manual and for high residue substances, tanks are generally pre-washed with discharge of washings to reception facilities.

Type C chemicals include: acetic acid; benzene; creosote (coal tar); and ferric chloride solution.

The cargo pumps for type C substances in chemical tankers built after 1st July 1986 must be capable of clearing 'water' from the tanks such that remains do not exceed 0.3 cubic metres (0.9 cubic metres for older ships). Guidance for the discharge of category C substances is obtained from the Procedures and Arrangements manual. These regulations are similar to those for type B substances.

Type D chemicals include: calcium chloride solution; calcium hydroxide solution; castor oil; and hydrochloric acid.

The discharge of type D chemicals into the sea is not permitted unless:

1 The ship is proceeding at not less than seven knots,
2 Content of the discharge is made up of only one part of the substance with ten parts of water,
3 The vessel is more than twelve miles from land.

There is a limit imposed on the quantity discharged, if the vessel is in a special area.

GARBAGE DISPOSAL

Domestic refuse ashore is completely removed and disposed of by burying or burning. Rubbish from ships has traditionally been dumped at sea where most of it sank or in the case of food wastes, was eaten by seabirds or fish. The introduction of plastic containers and packaging, use of synthetics for ropes and fishing nets, and the proliferation of plastic bags has made casual dumping a major nuisance. Research has shown that the majority of garbage washed up on British beaches, originated from ships. The plastic items do not rot or break down. They are also obvious and unsightly. The nuisance value extends to blockage of sea water inlet strainers and components such as ejectors. Oddments made of plastic, metal and other materials have proved dangerous to wild life.

Annex V of the IMO Marpol 73/78 convention which seeks to control disposal of garbage is now in force. These new regulations seek to reduce the garbage nuisance by imposing limits on the dumping of the various kinds of waste. The United Kingdom Regulations, which give effect to the IMO convention, came into force on the 31st December 1988 as The Merchant Shipping (Prevention of Pollution by Garbage) Regulations 1988.

There is a complete ban on the dumping of plastics at sea in any area. Options for disposal of plastics in the form of bags, packaging, synthetic ropes, synthetic fishing nets and any other substances which could be so categorized, include incinerating or on board retention until the vessel reaches a port with reception facilities. An incinerator implies air pollution together with the penalty of initial expenditure and operating costs (fuel). Legislation in the United Kingdom, requires provision of garbage disposal facilities by ports and terminals at a reasonable charge. Not all signatory countries will make this provision in the immediate future.

The dumping of refuse within 3 miles of any coastline is prohibited. Outside of this area, food wastes and other items such as paper products, rags, glass/bottles, crockery and metals can be disposed of, provided that they have been passed through a comminuter or grinder. Substances passing through a grinder must be rendered small enough to pass through a 25 mm screen.

Beyond the 12 miles line, comminution of the above refuse is not necessary. There is a 25 mile limit beyond which dunnage, lining and packing materials which may float, can be disposed of. Dumping of plastics is not allowed in any area. At a later date, controls for special areas which include the Mediterranean, Baltic and Black seas, together with other areas severely at risk from pollution by garbage, will be made more stringent. Timing of the further regulations is dependent on improvement of shore disposal facilities. When this happens, there will be a complete ban on the dumping of general refuse in the special areas. An exception made for food wastes will permit disposal only beyond the 12 mile limit.

References

The Merchant Shipping (Prevention of Oil Pollution) Regulations 1983. HMSO.
The Merchant Shipping (Prevention of Pollution by Garbage) Regulations 1988. HMSO.
The Merchant Shipping (Reception Facilities for Garbage) Regulations 1988. HMSO.
Statutory Instruments 1987 No. 551. Marine Pollution. The Merchant Shipping (Control of Pollution by Noxious Liquid Substances in Bulk) Regulation 1987.

CHAPTER 10

Production of Water

Modern low pressure evaporators and reverse osmosis systems give relatively trouble free operation particularly in comparison with the types that were fitted in older ships. They are sufficiently reliable to provide the water needed for the engine room and domestic consumption during continuous and unattended operation. Storage capacity for water can be donated to commercial earning. An advantage of low pressure evaporators is that they enable otherwise wasted heat from diesel engine jacket cooling water to be put to good use.

Reverse osmosis systems were installed to give instant water production capacity without extensive modifications (as with vessels commandeered for hostilities in the Falklands war). They are used to advantage on some passenger cruise vessels and are fitted in ships which may remain stopped at sea for various reasons (tankers awaiting orders, outside a 20 mile limit).

Warning is given in M Notice M620 that evaporators must not be operated within 20 miles of a coastline and that this distance should be greater in some circumstances. Pollution is present in inshore waters from sewage outfalls, disposal of chemical wastes from industry, drainage of fertilizers from the land and isolated cases of pollution from grounding or collision of ships and spillage of cargo.

LOW PRESSURE EVAPORATORS

Low pressure evaporators for the production of water may be heated by waste steam on steamships or by engine cooling water on motorships. The relatively low temperature jacket water entering at about 65°C and leaving at about 60°C will produce evaporation because vacuum conditions reduce boiling temperature of the sea water from 100°C to less than 45°C.

The single effect, high vacuum, submerged tube evaporator shown (Fig. 112) is supplied with diesel engine cooling water as the heating medium. Vapour evolved at a very rapid rate by boiling of the sea water feed tends to carry with it droplets of salt water which must be removed to avoid contamination of the product. The demister of knitted monel metal wire or polypropylene collects the salt filled water droplets as they are carried through by the air. These coalesce forming drops large enough to fall back against the vapour flow.

Evaporation of part of the sea water leaves a brine the density of which must be controlled by continual removal through a brine ejector or pump. Air and other gases released by heating of the sea water, but which will not condense, are removed by the air ejector. One of the gases liberated is CO_2 from calcium bicarbonate in the sea water. Loss of carbon dioxide from calcium bicarbonate leaves plain calcium carbonate which has poor solubility and a tendency to form soft, white scale. Other potential scale-forming salts are calcium sulphate and magnesium compounds.

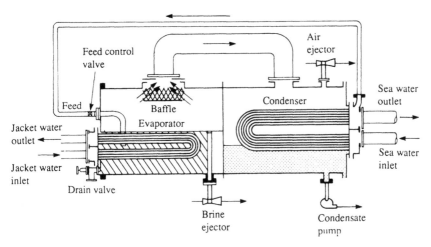

Fig 112   Evaporator

Scale is not a major problem where submerged heating coils reach a temperature of only 60°C. This heat is too low for formation of magnesium scales and provided brine density is controlled, calcium sulphate will not cause problems. Continuous removal of the brine by the brine pump or ejector, limits density. Another safeguard is the distillate return loop system, to cope with excess evaporation due to clean tube conditions or higher than usual heating water temperature. The higher evaporation rate would increase brine density and produce scale formation but output is restricted to the rated figure; any excess distillate being returned to the brine sump, to keep density down.

The small quantity of soft calcium carbonate scale can be removed by periodic cleaning with a commercially available agent or the evaporator can be continually dosed with synthetic polymer to bind the scale forming salt into a 'flocc' which mostly discharges with the brine. Use of continuous treatment will defer acid cleaning to make it an annual exercise. Without continuous treatment, cleaning may be necessary after perhaps two months.

Steam heated evaporators with their higher heating surface temperature, benefit more from chemical dosing, because magnesium hydroxide scales form when surfaces are at 80°C or more.

CORROSION

The steel shell of evaporators is prone to corrosion. Protection is provided in the form of natural rubber, rolled and bonded to the previously shot blasted steel. The adhesive is heat cured and integrity of the rubber checked by spark test.

DISTILLATE TREATMENT

The low operating temperature of the evaporator described, is not sufficient to sterilize the product. Despite precautions near the coast, harmful organisms

145

may enter with the sea water and pass through to the domestic water tank and system. Additionally there is a likelihood that while in the domestic tank, water may become infested due to build up of a colony of organisms from some initial contamination. Sterilization by addition of chlorine, is recommended in Merchant Shipping Notice M 1214. A later notice, M1401, states that the Electro-Katadyn process in use since the 1960s, has also been approved.

Distilled water having left behind the compounds previously dissolved in it, tastes flat and tends to be slightly acidic due to ready absorption of carbon dioxide (CO_2). This condition makes it corrosive to pipe systems and less than beneficial to the human digestive tract.

TREATMENT WITH CHLORINE STERILIZATION

Initial treatment involves passing the distillate through a neutralite unit, containing magnesium and calcium carbonate. Some absorption of carbon dioxide from the water and the neutralizing effect of these compounds, removes acidity. The addition of hardness salts also gives the water a better taste. Chlorine, being a gas, is injected for sterilizing purposes, as a constituent of sodium hypochlorite (a liquid) or as solid granules of calcium chloride dissolved in water. The addition is set to bring chlorine content to 0.2 p.p.m.

While the water resides in the domestic tank, chlorine should preserve sterility. In the long term, it will evaporate.

The passage of water from storage tanks to the domestic system is by way of a carbon filter which removes the chlorine taste.

TREATMENT WITH SILVER ION ADDITION

The electrokatadyn process (Fig. 113) accepted as an alternative to chlorination (see M 1401) involves the use of a driven silver anode to inject silver ions into the distilled water product of the low temperature evaporator. Silver is toxic to the various risk organisms. Unlike the gas chlorine, it will not evaporate but remains suspended in the water.

The sterilizer is placed close to the production equipment with the conditioning unit being installed after the sterilizer and before the storage tank.

The amount of metal released to water passing through the unit is controlled by the current setting. If a large volume has to be treated only part is by passed through and a high current setting is used to inject a large amount of silver. The bypassed water is then added to the rest in the pipeline. With low water flow, all of the water is delivered through the device and the current setting is such as to give a concentration of 0.1 p.p.m. of silver. Silver content of the system should be 0.08 p.p.m. maximum.

ULTRA-VIOLET LIGHT

Chlorine and silver ion sterilization give lasting protection but sterilization by ultra-violet light, although instantly effective, will not last and will not, therefore, prevent re-infection. It is used in conjunction with other methods but not as the sole means of sterilization. Ultra-violet radiation from low pressure mercury lamps is used for pretreatment disinfection in some reverse osmosis plant.

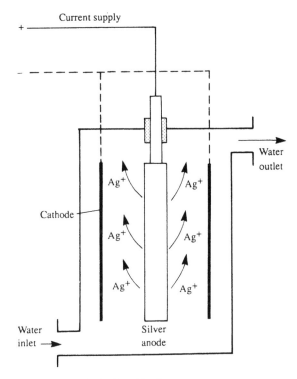

Fig. 113   Electrokatadyn method of sterilization

REVERSE OSMOSIS

Osmosis is the term used to describe the passage of pure water from one side of a semi-permeable membrane into a salt or other solution on the other, with the result that the salt solution is diluted but the water remains pure. The membrane acts as a filter allowing passage of water but not of the salt. The action will continue despite rise of head of the salt solution relative to that of the pure water. A more dramatic laboratory demonstration uses a parchment-covered, inverted thistle funnel partly filled with solution and immersed in a container of pure water (Fig. 114). The liquid level in the funnel rises as pure water passes through the parchment and into the solution. Osmotic pressure can be obtained by measuring the head of the solution when the action ceases.

The membrane and the parchment are semi-permeable and allow the water molecules through but not the larger salt molecules. The phenomenon is important in the absorption of water through the roots of plants and in animal and plant systems generally.

Reverse osmosis is a water filtration process which makes use of semi-permeable membrane materials. Salt water on one side of the membrane is pressurized by a pump and forced against the material. Pure water passes through, but not the salts (Fig. 115). For production of large amounts of pure

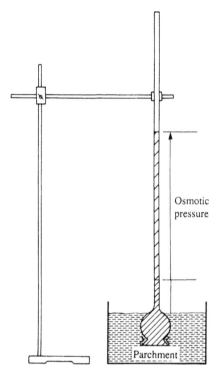

Fig 114   Demonstration of osmosis

water, the membrane area must be large and it must be tough enough to withstand the pump pressure. The material used for sea water purification is spirally wound polyamide or polysulphonate sheets. One problem with any filtration system is that deposit accumulates and gradually blocks the filter. Design of the cartridges (Fig. 116) is therefore such that the sea water feed passes over the membrane sheets so that the washing action keeps the surfaces clear of deposit. A dosing chemical is also injected to assist the action.

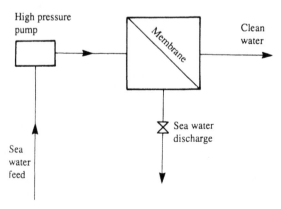

Fig. 115   Reverse osmosis

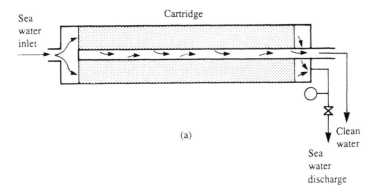

(a)

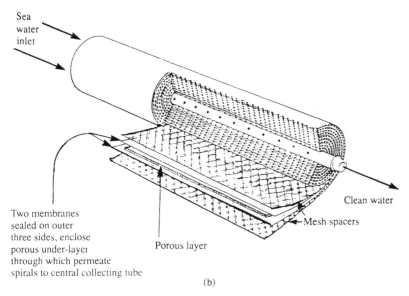

(b)

Fig. 116   Reverse osmosis cartridge

The pump delivery pressure for a reverse osmosis system, of 60 bar (900 lb/in²) calls for a robust reciprocating or gear pump. The system must be protected by a relief arrangement.

PRETREATMENT AND POST TREATMENT

Sea water feed for reverse osmosis plant is pretreated before being passed through. The chemical sodium heta mexa phosphate is added to assist wash through of salt deposit on the surfaces of the elements and the sea water is sterilized to remove bacteria which could otherwise become resident in the filter. Chlorine is reduced by the compressed carbon filter while solids are removed by the other filters.

Treatment is also necessary to make the water product of reverse osmosis potable. The method is much the same as for water produced in low temperature evaporators.

TREATMENT OF WATER FROM SHORE SOURCES

There is a risk that water supplied from ashore may contain harmful organisms which can multiply and infect drinking or washing water storage tanks. All water from ashore, whether for drinking or washing purposes, is to be sterilized. When chlorine is used, the dose must be such as to give a concentration of 0.2 p.p.m. The Department of Transport recommends in Merchant Shipping Notice number M 1214 that because of the risk from *legionella* bacteria entering the respiratory system by way of fine mist from a shower spray, all water including that for washing only, should be treated by sterilization.

The transfer hose for fresh water is to be marked and kept exclusively for that purpose. The ends must be capped after use. The hose must be stored clear of the deck where contamination is unlikely.

DOMESTIC WATER TANKS

Harmful organisms in drinking water storage tanks have caused major health problems on passenger vessels and to oil platform personnel, among others. Tanks should, at six-month intervals, be pumped out and, if necessary, the surfaces hosed to clean them. At the twelve-month inspection, cleaning and recoating may be needed. Washing with a 50 p.p.m. solution of chlorine is suggested. Super-chlorinating when the vessel is drydocked, consists of leaving a 50 p.p.m. chlorine solution in the tank over four hours, followed by flushing.

Tanks surfaces are prepared by wire brushing and priming, before application of a cement wash. Epoxy and other coatings developed for use in fresh water tanks, are available.

References

M 1214 Recommendations to Prevent Contamination of Ships' Fresh Water Storage and Distribution Systems.

M 1401 Disinfection of Ships' Domestic Fresh Water.

The Merchant Shipping (Crew Accommodation) Regulations 1978. HMSO.

Allanson, J. T. and Charnley, R. (1987). Drinking Water from the Sea: Reverse Osmosis, the Modern Alternative. *Trans. I. Mar. E.,* vol. 95, paper 38.

Gilchrist, A. (1976). Sea Water Distillers. *Trans. I. Mar. E.,* vol. 88.

Hill, E. C. (1987). Legionella and Ships' Water Systems. MER.

CHAPTER 11

Noise and Vibration

NOISE

Main machinery space noise in ships with steam auxiliaries and powered by steam reciprocating machinery or slow speed two stroke diesels was moderate. Spoken communication was possible at only a little above normal conversational levels except in the area of the propeller shaft tunnel. Shouted conversation was possible in ships with medium speed diesel driven generators and steam turbine or diesel main propulsion until the advent of turbo-charging. However, noise in the steering gear and tunnel areas was often excessive. The placing of engines aft, brought more extreme noise and vibration in this area.

The advent of turbo-charging brought a tremendous increase in machinery space noise. The problem was exacerbated by a tendency to run medium-speed, four-stroke engines at higher speeds to obtain higher powers.

Sound levels from turbo-chargers, medium-speed engines and the propeller shafting has been augmented severely by the noise from some hydraulic power installations. The general increase in amount of equipment installed and the large number of electric motors fitted has also added to the problem.

High noise levels have seriously interfered with communication in the machinery space. This causes major problems when working or handing over responsibility.

In an emergency, the situation is made much more serious if verbal communication is impossible.

The degree of noise in some machinery spaces is powerful enough to interfere with clear thinking and decision making.

At a personal level, exposure to high sound energy has caused ear damage and permanent deafness to many people. A study carried out on 184 Scandinavian ships among engineering staff revealed that 44 per cent had suffered grave hearing loss, 42 per cent some hearing loss and only 14 per cent had normal hearing.

Guidelines have now been set by various national authorities and the United Kingdom has issued a code of practice for noise levels in ships which sets recommended maximums for various areas. The Code gives advice on measurement and instructions on the method for carrying out a noise survey. This procedure should be carried out on new and existing vessels.

Noise in continuously manned machinery spaces, workshops and stores (all classed as machinery spaces 'A') should be below the recommended limit of 90 dB(A) or ISO NR 85.

Unmanned areas are categorized as machinery spaces 'B' and for these the recommended limit is 110 dB(A) or ISO NR 105.

The limit for machinery and cargo control rooms which are occupied for normal operations is set at 75 dB(A) or ISO NR 70. This particular maximum was chosen as one which would not interfere with clear thinking and such as to permit adequate speech communication. A note in the code of practice states: 'Face-to-face conversation should be satisfactory at a distance of approximately 0.75 m using a raised voice. Telephone usage is likely to be slightly difficult'.

The recommended maximum noise level in the wheelhouse is 65 dB(A). A note relating to this states: 'Face-to-face conversation should be satisfactory at distances of up to 1.2 m for normal voice effort and up to about 9 m when shouting. Telephone and radio usage should be acceptable'.

Other recommended maximum levels are:

Galleys and pantries—70 dB(A)
Sleeping cabins and hospital—60 dB(A)
Day cabins and offices—65 dB(A)
Mess rooms and recreation rooms—65 dB(A)

Damage to the hearing is based on sound level and time of exposure. There are limits set on the safe time of exposure to certain levels of sound.

Hearing is most at risk in machinery spaces where noise levels may be greater than the recommended maximum. A noise level, for example, has been measured in a location between two high speed 1800 r.p.m. diesel generators at 120 dB(A). Exposure to this degree of sound energy is potentially harmful, and the acceptable maximum daily dose if the ears are not protected is 30 seconds.

Unprotected exposure to a noise level of 110 dB(A), as measured in a small machinery space with medium speed and fast diesel units, should be limited to 5 minutes. The figure of 110 dB(A) is the recommended limit for unmanned machinery spaces in which personnel may be engaged in maintenance work on a daily basis!

Exposure time should be limited to 50 minutes per day, for readings of 100 dB(A) Readings of 105 have been taken near slow-speed engine cylinder heads and figures of 100 dB(A) between medium-speed diesel generators.

The ideal limit for manned machinery spaces of 90 dB(A) should be endured for not longer than 8 hours per day. The limit for 87 dB(A) is 16 hours. Only when the noise is at 85 dB(A) or less is there no limit on exposure time.

WARNING

Ear protection is necessary and a suitable warning should be posted at the entrance to any space in which noise levels reach or exceed 90 dB(A).

Many of the available ear muffs and plugs are not adequate. The performance of approved types should be checked against measured noise levels to ensure that they are capable of reducing effects to an equivalent of 85 dB(A) (for unlimited exposure time), 87 dB(A) (for 16 hours) or 90 dB(A) (for 8 hours).

VIBRATION AND CONDITION MONITORING

Routine maintenance of ships machinery has been based on running hours as recommended by equipment manufacturers. Overhaul times have then been adjusted by experience and the requirements of government department and/or classification society survey. The operating and maintenance experience of ships' engineers provided the foundation for planned maintenance schemes. Unfortunately, planned maintenance decrees exact periods between overhauls and tends to remove flexibility and sensible alteration of running hours, by engine room staff. The result is that planned maintenance has proved to be not cost effective.

Equipment has been frequently taken out of service for planned maintenance and found to have no faults and a potential for many more hours of operation. The cost of stripping down is paid not only in terms of replacement joints, seals and parts but sometimes in damage inflicted and ultimate breakdown built in, during the procedure of opening the equipment.

Dismantling machinery and equipment for survey may be costly in the same way.

The alternative to planned maintenance, of repairing after breakdown, can be expensive in terms of the extent of resulting damage. Breakdown maintenance is not the remedy of the prudent engineer.

Condition monitoring, another and more acceptable option, is becoming more widely used as the means of determining when machinery should be overhauled.

Classification societies are now willing to accept, by mutual agreement, vibration monitoring as an alternative to taking machinery apart for survey. Used oil analysis an alternative method of determining engine condition, is seen as another surveillance method.

VIBRATION ANALYSIS

Noticeable or at least measureable vibration, is associated with most mechanical problems that occur in machines. A rotating machine with inadequate foundations or bearing support, imbalance or misalignment may suffer severe and obvious vibration. Other defects which would be made apparent by increasing vibration, are damaged or worn bearings and gear teeth which are mating badly, or are worn or damaged. Varying degrees of vibration, depending on the progress of the fault, will result from these and other defects. There are also those unavoidable vibrations associated with certain equipment (i.e., reciprocating machinery) where there are unbalanced forces.

Measurements of vibration can be made: (a) if a problem develops and diagnosis is necessary; (b) as a means of condition monitoring for maintenance; and (c) as a substitute for classification society survey where agreement has been reached with the society.

VIBRATION MONITORING INSTRUMENT READINGS

Vibration readings are taken mainly on the bearing housings of rotating machinery as close as possible to the shaft. The vibration pick-up is placed on each bearing in turn to record in the vertical, horizontal and axial directions. Readings may also be taken on casings, supports and at other relevant points. Readings of vibration magnitude and frequency are recorded manually from basic instruments but some equipment incorporates data collectors from which information is fed to a computer for analysis. Instruments with a built-in capability for on site analysis are available. There are also fixed installations with a display unit.

FAULT DIAGNOSIS

An investigation of excessive vibration relies mainly on vibration frequency readings to identify normal and abnormal vibrations. The readings are obtained with the machinery in a steady running state. Other factors which will contribute perhaps in a major way to the identification of problems are also taken into account. Thus details such as speed, load and operating temperatures are noted together with any history of component failure.

CONDITION MONITORING AND PLANNED MAINTENANCE

The shortcomings of a planned maintenance scheme, based on calendar or running hours to dictate when equipment should be opened for inspection and overhaul, can be reduced by incorporation of condition monitoring. Many shipping companies now use a combination of the two methods to provide greater cost effectiveness. The condition and performance is checked on a planned basis, but items are only opened for examination when vibration and other readings show there has been a deterioration.

SURVEY OF MACHINERY

To satisfy the requirements of classification, the condition of ships and their machinery has to be demonstrated at regular intervals to surveyors from the societies. A complete survey of a ship's machinery can be carried out at four-yearly intervals or the alternative of a continuous survey over a period of five years can be agreed on. The latter continuous survey of machinery has involved the opening up and examination of approximately one-fifth of the machinery in each year.

In the past, the classification societies have permitted designated chief engineers to carry out surveys themselves in ports and places where no surveyor from the society was available. This facility has now been extended. Designated chief engineers are now empowered to carry out surveys at sea or in any port on much but not all of the machinery. A condition that must be satisfied is that the vessel has a suitable planned maintenance scheme in operation with, possibly, condition monitoring. Machinery which is operated with condition monitoring has only to be opened for examination when readings indicate a deterioration.

References

Approved Planned Maintenance Schemes an Alternative to C.S.M. Lloyds Regulations.

Code of Practice for Noise Levels in Ships. HMSO.

Flising, A. (1978). Noise Reduction in Ships. *Trans. I. Mar. E.*, vol. 90.

Noise Levels on Board Ships. IMO.

Thomas, B. (1990). Vibration—identifying the source, solving the problem. MER.

Index